Lecture Notes in Mechanical Engineering

Lecture Notes in Mechanical Engineering (LNME) publishes the latest developments in Mechanical Engineering—quickly, informally and with high quality. Original research reported in proceedings and post-proceedings represents the core of LNME. Volumes published in LNME embrace all aspects, subfields and new challenges of mechanical engineering. Topics in the series include:

- Engineering Design
- Machinery and Machine Elements
- Mechanical Structures and Stress Analysis
- Automotive Engineering
- Engine Technology
- Aerospace Technology and Astronautics
- Nanotechnology and Microengineering
- Control, Robotics, Mechatronics
- MEMS
- Theoretical and Applied Mechanics
- Dynamical Systems, Control
- Fluid Mechanics
- Engineering Thermodynamics, Heat and Mass Transfer
- Manufacturing
- Precision Engineering, Instrumentation, Measurement
- Materials Engineering
- Tribology and Surface Technology

To submit a proposal or request further information, please contact the Springer Editor of your location:

China: Ms. Ella Zhang at ella.zhang@springer.com
India: Priya Vyas at priya.vyas@springer.com
Rest of Asia, Australia, New Zealand: Swati Meherishi at swati.meherishi@springer.com
All other countries: Dr. Leontina Di Cecco at Leontina.dicecco@springer.com

To submit a proposal for a monograph, please check our Springer Tracts in Mechanical Engineering at https://link.springer.com/bookseries/11693 or contact Leontina.dicecco@springer.com

Indexed by SCOPUS. All books published in the series are submitted for consideration in Web of Science.

More information about this series at https://link.springer.com/bookseries/11236

Adam Hamrol · Marta Grabowska ·
Damjan Maletič
Editors

Advances in Manufacturing III

Volume 3 - Quality Engineering: Research and Technology Innovations, Industry 4.0

Springer

Editors
Adam Hamrol
Department of Production Engineering
Poznan University of Technology
Poznan, Poland

Marta Grabowska
Production Engineering Department
Poznan University of Technology
Poznan, Poland

Damjan Maletič
Faculty of Organizational Sciences
University of Maribor
Kranj, Slovenia

ISSN 2195-4356 ISSN 2195-4364 (electronic)
Lecture Notes in Mechanical Engineering
ISBN 978-3-031-00166-6 ISBN 978-3-031-00218-2 (eBook)
https://doi.org/10.1007/978-3-031-00218-2

This Springer imprint is published by the registered company Springer Nature Switzerland AG
The registered company address is: Gewerbestrasse 11, 6330 Cham, Switzerland

Preface

This volume of Lecture Notes in Mechanical Engineering gathers selected papers presented at the 7th International Scientific-Technical Conference MANUFACTURING 2022, held in Poznan, Poland, on May 16–19, 2022. The conference was organized by the Faculty of Mechanical Engineering, Poznan University of Technology, Poland.

The aim of the conference was to present the latest achievements in the broad field of mechanical engineering and to provide an occasion for discussion and exchange of views and opinions. The conference covered topics in:

- mechanical engineering
- production engineering
- quality engineering
- measurement and control systems
- biomedical engineering.

The organizers received 165 contributions from 23 countries around the world. After a thorough peer-review process, the committee accepted 91 papers for conference proceedings prepared by 264 authors from 23 countries (acceptance rate around 55%). Extended versions of selected best papers will be published in the following journals: *Management and Production Engineering Review, Bulletin of the Polish Academy of Sciences: Technical Sciences, Materials, Applied Sciences.*

The book **Advances in Manufacturing III** is organized into five volumes that correspond to the main conference disciplines mentioned above.

Advances in Manufacturing III - Volume 3 - Quality Engineering: Research and Technology Innovations, Industry 4.0 consists of articles regarding challenges of quality engineering and quality management related to the Fourth Industrial Revolution and humankind's pursuit for sustainable development. They show that climatic and environmental threats require a product design paradigm shift in order to meet the expectations and requirements of consumers. A range of subjects is covered, including risk analysis in production processes and in product life cycle. Some articles address the use of big data and quality management methods in production companies and in higher education or public sector. Some chapters are devoted to the implementation of statistical process control, vision control and inspection in the production processes. The articles significantly contribute to the development of quality engineering and quality management and are attractive for academic, business and industry. This book includes 20 chapters, prepared by 61 authors from ten countries.

We would like to thank the members of the international program committee for their hard work during the review process.

We acknowledge all people that contributed to the staging of MANUFACTURING 2022: authors, committees and sponsors. Their involvement and hard work were crucial to the success of the MANUFACTURING 2022 conference.

May 2022

Adam Hamrol
Marta Grabowska
Damjan Maletič

Organization

Steering Committee

General Chair

Adam Hamrol	Poznan University of Technology, Poland

Chairs

Olaf Ciszak	Poznan University of Technology, Poland
Magdalena Diering	Poznan University of Technology, Poland
Justyna Trojanowska	Poznan University of Technology, Poland

Scientific Committee

Ahmad Majdi Abdul-Rani	University of Technology PETRONAS, Malaysia
Dario Antonelli	Politecnico di Torino, Italy
Katarzyna Antosz	Rzeszow University of Technology, Poland
Jorge Bacca-Acosta	Fundación Universitaria Konrad Lorenz, Colombia
Zbigniew Banaszak	Koszalin University of Technology, Polska
Andre D. L. Batako	Liverpool John Moores University, UK
Stefan Berczyński	West Pomeranian University of Technology in Szczecin, Poland
Alain Bernard	Ecole Centrale de Nantes, France
Kristina Berladir	Sumy State University, Ukraine
Christopher Brown	Worcester Polytechnic Institute, USA
Anna Burduk	Wrocław University of Science and Technology, Poland
Marcin Butlewski	Poznan University of Technology, Poland
Lenka Čepová	VSB Technical University of Ostrava, Czech Republic
Nadežda Čuboňová	University of Žilina, Slovakia

Somnath Chattopadhyaya	Indian School of Mines, India
Olaf Ciszak	Poznań University of Technology, Polska
Jacek Diakun	Poznań University of Technology, Poland
Magdalena Diering	Poznan University of Technology, Poland
Ewa Dostatni	Poznan University of Technology, Poland
Jan Duda	Cracow University of Technology, Poland
Davor Dujak	J.J. Strossmayer University of Osijek, Croatia
Milan Edl	University of West Bohemia, Czech Republic
Brigita Gajšek	University of Maribor, Slovenia
Mosè Gallo	University of Naples Federico II, Italy
Adam Gałuszka	Silesian University of Technology, Poland
Józef Gawlik	Cracow University of Technology, Poland
Adam Gąska	Cracow University of Technology, Poland
Adam Górny	Poznan University of Technology, Polska
Filip Górski	Poznan University of Technology, Poland
Adam Hamrol	Poznan University of Technology, Poland
Mukund Harugade	Padmabhooshan Vasantraodada Patil Institute of Technology, India
Ivan Hudec	Slovak University of Technology in Bratislava, Slovakia
Jozef Husár	Technical University of Košice, Slovakia
Aminul Islam	Technical University of Denmark, Denmark
Vitalii Ivanov	Sumy State University, Ukraine
Andrzej Jardzioch	West Pomeranian University of Technology in Szczecin, Poland
Sabahudin Jašarević	University of Zenica, Bosnia and Herzegovina
Małgorzata Jasiulewicz-Kaczmarek	Politechnika Poznańska, Poland
Wojciech Kacalak	Technical University of Koszalin, Poland
Sławomir Kłos	University of Zielona Góra, Polska
Lucia Knapčíková	Technical University of Košice, Slovakia
Damian Krenczyk	Silesian University of Technology, Poland
Jolanta Krystek	Silesian University of Technology, Poland
Józef Kuczmaszewski	Lublin University of Technology, Poland
Agnieszka Kujawińska	Poznan University of Technology, Poland
Ivan Kuric	University of Žilina, Slovak Republic, Slovak Republic
Piotr Łebkowski	AGH, University of Science and Technology, Poland
Stanisław Legutko	Poznan University of Technology, Polska
Sławomir Luściński	Kielce University of Technology, Poland
José Machado	University of Minho, Portugal
Vijaya Kumar Manupati	National Institute of Technology, Warangal, India
Uthayakumar Marimuthu	Kalasalingam Academy of Research and Education, India

Michał Wieczorowski	Poznan University of Technology, Poland
Hanna Włodarkiewicz-Klimek	Poznan University of Technology, Poland
Szymon Wojciechowski	Poznan University of Technology, Poland
Ralf Woll	BTU Cottbus-Senftenberg, Germany
Jozef Zajac	Košice University of Technology, Slovakia
Nermina Zaimovic-Uzunovic	University of Zenica, Bosnia, and Herzegovina

Program Committee

Karla Alvarado-Ramírez
José Angel
Dario Antonelli
Katarzyna Antosz
Erfan Babaee Tirkolaee
Jorge Bacca-Acosta
Diana Baila
Prashanth Bandari
Lucía Barcos-Redín
Andre Batako
Petr Beneš
Kristina Berladir
Marcin Białek
Mikołaj Bilski
Matej Borovinšek
Anna Borucka
Cristina Borzan
Sara Bragança
Daniel Brissaud
Christopher Brown
Paweł Buń
Anna Burduk
Bartłomiej Burlaga
Jacek Buśkiewicz
Marcin Butlewski
Massimiliana Carello
Fernando Castillo
Robert Cep
Miroslav Cisar
Olaf Ciszak
Felipe Contreras
Tomáš Coranič
Eric Costa
Margareta Coteata
Dorota Czarnecka-Komorowska
Reggie Davidrajuh
Mario del Valle
Boris Delibašić
Jakub Demčák
Jacek Diakun
Magdalena Diering
Grzegorz Domek
Ewa Dostatni
Radosław Drelich
Jan Duda
Luboslav Dulina
Eduard Franas
Paweł Fritzkowski
Brigita Gajšek
Mosè Gallo
Bartosz Gapiński
Katarzyna Gawdzińska
Adam Gąska
Dominik Gojdan
Arkadiusz Gola
Fernando González-Aleu
Adam Górny
Filip Górski
Cezary Grabowik
Marta Grabowska
Jakub Grabski
Damian Grajewski
Patrik Grznár
Aleksander Gwiazda
Michal Hajžman
Adam Hamrol
Mukund Harugade
Amalija Horvatić-Novak

Stella Hrehová
Jozef Husár
Vitalii Ivanow
Carmen Jaca
Rajat Jain
Michał Jakubowicz
Małgorzata Jankowska
Andrzej Jardzioch
Małgorzata Jasiulewicz-Kaczmarek
Anna Karwasz
Jakub Kaščak
Sławomir Kłos
Lucia Knapčíková
Adam Koliński
Elif Kongar
Boris Kostov
Kateryna Kovbasiuk
Arkadiusz Kowalski
Tomasz Kowaluk
Damian Krenczyk
Grzegorz Królczyk
Józef Kuczmaszewski
Agnieszka Kujawińska
Panagiotis Kyratsis
Georgios Lampropoulos
Peter Lazorík
Stanislaw Legutko
Andrzej Loska
Czesław Łukianowicz
José Machado
Marek Macko
Ján Majerník
Vitalii Maksymenko
Damjan Maletič
Matjaž Maletič
Vijaya Kumar Manupati
Sven Maricic
Jorge Martín Gutiérrez
Thomas Mathia
Maciej Matuszewski
Dariusz Mazurkiewicz
Jakub Michalski
Dariusz Mikołajewski
Janusz Mleczko
Ladislav Morovic
Keddam Mourad
Leticia Neira
Zdeněk Neusser
Magdalena Niemczewska-Wójcik
Przemysław Niewiadomski
Czesław Niżankowski
Nejc Novak
Filip Osiński
Erika Ottaviano
Eren Ozceylan
Ancuta Pacurar
Kyratsis Panagiotis
Jason Papathanasiou
Waldemar Paszkowski
Justyna Patalas-Maliszewska
Ivan Pavlenko
Paweł Pawlus
Jarosław Pempera
Dragan Peraković
Alejandro Pereira
Dariusz Plinta
Florin Popister
Evangelos Psomas
Paulina Rewers
Francisco Gabriel Rodriguez-González
Michał Rogalewicz
Izabela Rojek
Mirosław Rucki
Biserka Runje
Dominik Rybarczyk
Michał Rychlik
Milan Saga
Alžbeta Sapietova
Filip Sarbinowski
Holger Schlegel
Armin Schleinitz
Dariusz Sędziak
Robert Sika
Zbyněk Šika
Varinder Singh
Bożena Skołud
António Lucas Soares
Ewa Stachowska
Dorota Stadnicka
Sergiu Dan Stan

Roman Starosta
Beata Starzyńska
Krzysztof Stępień
Tomasz Stręk
Manuel Suárez-Barraza
Marcin Suszynski
Grażyna Sypniewska-Kamińska
Marek Szostak
Jiří Tengler
Samala Thirupathi
Francisco Torres
Justyna Trojanowska
Piotr Trojanowski
Krzysztof Tyszczuk
M. Uthayakumar
Leonilde Varela
Nikola Vitkovic
Adrián Vodilka
Sachin Waigaonkar
Tomasz Walczak
Łukasz Warguła
Gerhard-Wilhelm Weber
Katarzyna Węgrzyn-Wolska
Radosław Wichniarek
Michał Wieczorowski
Dariusz Więcek
Hanna Włodarkiewicz-Klimek
Szymon Wojciechowski
Adam Woźniak
Ryszard Wyczółkowski
Nermina Zaimovic-Uzunovic
Ivan Zajačko
Magdalena Żukowska
Krzysztof Żywicki

Special Sessions

Manufacturing and Management Approaches and Tools for Companies in Era of Digitalization

Special Session Organizing Committee

Leonilde Varela	University of Minho, Portugal
Justyna Trojanowska	Poznan University of Technology, Poland
Vijaya Kumar Manupati	Mechanical Engineering Department, India
Paulina Rewers	Poznan University of Technology, Poland

Operational Innovation and Excellence in Organizations of Twenty-First Century. Lean Thinking, Kaizen Philosophy, and Toyota Production System

Manuel F. Suárez-Barraza	Universidad de las Américas Puebla, Mexico
Manuel F. Morales-Contreras	Universidad Pontificia Comillas, Spain
Agnieszka Kujawińska	Poznan University of Technology, Poland
Marta Grabowska	Poznan University of Technology, Poland

Cyber-Physical Production Systems: From Design to Applications

Katarzyna Antosz	Rzeszow University of Technology, Poland
José Machado	University of Minho, Portugal
Erika Ottaviano	University of Cassino and Southern Lazio, Italy
Pierluigi Rea	University of Cagliari, Italy

Augmented, Virtual and Mixed Reality Systems

Jozef Husár	Technical University of Košice, Slovak Republic
Jakub Kaščak	Technical University of Košice, Slovak Republic
Paweł Buń	Poznan University of Technology, Poland

Intelligent Methods Supporting Manufacturing Systems Efficiency

Anna Burduk	Wroclaw University of Science and Technology, Poland
Andre Batako	Liverpool John Moores University, UK
Dorota więcek	University of Bielsko-Biala, Poland
Ivan Kuric	University of Žilina, Slovak Republic

Sustainability on Production in the Aspect of Industry 4.0

Ewa Dostatni	Poznan University of Technology, Poland
Izabela Rojek	Kazimierz Wielki University in Bydgoszcz, Poland
Jacek Diakun	Poznan University of Technology, Poland
Dariusz Mikołajewski	Kazimierz Wielki University in Bydgoszcz, Poland

Design and Rapid Manufacturing of Customized Medical Products

Filip Górski	Poznan University of Technology, Poland
Magdalena Żukowska	Poznan University of Technology, Poland
Razvan Pacurar	Technical University of Cluj-Napoca, Romania

Metrology and Statistical Analysis of Measurement and Control Systems

Alejandro Pereira Domínguez	University of Vigo, Spain
Lenka Čepová	VSB—Technical University of Ostrava, Czech Republic
Magdalena Diering	Poznan University of Technology, Poland
Bartosz Gapiński	Poznan University of Technology, Poland

Analytical and Computational Methods in (Bio-) Mechanical and Material Engineering

Tomasz Stręk	Poznan University of Technology, Poland
Roman Starosta	Poznan University of Technology, Poland
Nejc Novak	University of Maribor, Slovenia
Pavel Polach	University of West Bohemia, Czech Republic

Contents

Risk Analysis as a Way to Improve the Efficiency of Production Processes – Case Study

Dominika Rysińska-Wojtasik(✉) and Artem Balashov

Faculty of Mechanical Engineering, Wroclaw University of Science and Technology, Wrocław, Poland
{dominika.rysinska-wojtasik,artem.balashov}@pwr.edu.pl

Abstract. The importance of risk management in enterprises is constantly increasing, because it has influence on the company's leading positions, increasing its competitiveness in markets. It also helps to increase the efficiency of fundamental production processes by reduction of production losses. At present time, there is no unified assessment methodology of risk impact of the enterprise efficiency activity. Each enterprise uses independently developed methods for calculating the risk. The paper describes a methodology for analysing and assessing the risk in production system. The use of integrated methodology of risk assessment and analysis has been discussed based on case study from automotive company. The author describes the matter of performing identification of possible risks from 4M group (machine, material, method, man), their continuous analysis, assessment and mitigation method. This article features a case study by which production efficiency disturbing factors have been identified. Additionally, the risk levels have been assessed and process improving actions indicated.

Keywords: Risk · Risk assessment · Risk analysis · Risk management · Production process · Continuous improvement · 4M method

1 Introduction

The aim of contemporary enterprises is to constantly improve the areas of their activities. In order to do that, they use various methods and tools meant for identifying problems, weaknesses and the steps needed to be made to minimize or eliminate the possibility of their occurrence in the system [1, 2]. The issue of risk management tackles such actions. Proper, systematic and structured approach to risk management contributes to reduction of risks factors and, at the same time, it gives tangible benefits in the form of improving the effectiveness of all processes.

Managers seek to improve those areas of the production system where the strategic importance for the organization is essential and where the level of risk is the highest [3, 4]. In this dissertation, it has been assumed that the production losses are determinants of level of risk value because they reflect that the production processes in the system failed to accomplish assumed goals.

The areas of production systems that have been analyzed are in the first place related to the processes performed in the system. The input factors of those processes can be

A. Hamrol et al. (Eds.): MANUFACTURING 2022, LNME, pp. 1–11, 2022.
https://doi.org/10.1007/978-3-031-00218-2_1

divided into four main categories: materials, machines, methods and men (4M) [5]. It is thought that the concept of 4M is the foundation of functioning of every company, hence it is impossible to optimize any kind of process without stabilizing the 4M area first. For this reason, this dissertation has taken the 4M area as the main factor of potential source of losses [6, 7].

Risk factors in these areas disrupt and obstruct the fulfillment of assumed goals in the system which impacts the effectiveness of production processes and reduces it. To reduce negative effect of risk factors in production systems, the company needs reliable and appropriate risk management and the introduction of an effective and comprehensive risk management methodology.

2 Literature Review

The methods used for analysis and risk assessment help enterprises to interpret the threats in the environment – risk factors resulting in failure to achieve goals and, furthermore, decrease in effectiveness of production systems. There are several dozens of methods of risk assessment and analysis used in theory and practice [8–12]. The most extensive classification of tools and methods for risk assessment and analysis is presented in EN 31010:2010 standard which specifies as much as thirty-one methods being used in economic practice [13].

In relation to their character, those methods can be divided into quantitative, qualitative and mixed. Although the aim is common - to minimize the risk - these methods present a broad spectrum of usage, ranging from expert reflection to methods based on creating schematics and logical event scenarios. Table 1 presents the list from EN 31010:2010 standard related to the possibility of applying chosen method in risk assessment and analysis.

Table 1. Selected methods of risk assessment and analysis.

Method	Risk identification	Risk analysis			Risk assessment
		Severity	Occurrence	Detection	
FMEA analysis	++	++	++	++	++
Tree decision model	–	++	++	+	+
Monte carlo	–	–	–	–	++
Cause/effect matrix	++	++	++	++	+
Check list	++	++	++	++	++
If-What analysis	++	++	++	++	++
HAZOP analysis	++	++	+	+	+
FTA	+	–	++	+	+
ETA	+	++	+	+	–
HRA	++	++	++	++	+
LOPA	+	++	+	+	–
Description: – not used + used ++ commonly used					

Comparing the collected data, it can be noticed that some methods have limitations in relation to their usage in contemporary production systems. Presented methods are not complex enough which means that they do not always contain identification of threats, their analysis or risk assessment.

With reference to the conducted overview [14, 15], similar conclusions can be made. The methods presented in literature focus on individual cases of managing given risk factor [16–19]. In order to assess the risk comprehensively, it is necessary to modify the method in order to enable its application. The analysis and value of risk level in each method should refer to the whole production system [20–22], rather than to individual issues, which would prove that the method is complex and universal. Also, the methods of analysis and risk assessment available in the market do not ensure the possibility of analysing risk factors in view of the sources of their occurrence. Moreover, these methods do not enable viewing those risk factors with reference to their categories and their value in each group.

3 Methodology of Risk Assessment

Risk management facilitates the risk assessment of the production process by recognition and then grouping the risk factors. The presented risk assessment methodology uses the fact of dividing risk factors into **4M categories (machine (1M), material (2M), method (3M), man (4M))** [23]. The 4M approach considers all the main sources of risk [24].

Mentioned risk factors have influences on planned realization of production plan, what is closely connected with processes effectiveness. The presented methodology refers to FMEA analysis - it allows to find problems in the organization and creates a space for improvement. The methodology divides the risk assessment related to the presence of risk factors into the following stages [25]:

STAGE 1: Identification of risk factors in the process by usage of standardized data collecting tools to find the risk.
STAGE 2: Categorization of risk factors - risk factors distribution into 4M categories. Each single case should be assigned to the appropriate category.
STAGE 3: Analysis of risk factors from each category including the frequency of **occurrence (O), severity** of their impact **(S)** and probability of their **detection (D)** – based on proposed evaluation criteria.
STAGE 4: Assessment of risk level:

- **Single Risk Indicator** (**SRI**) regarding 4M categories - (SRI1M, SRI2M, SRI3M, SRI4M),
- **Risk Level Indicator** (**RLI**) regarding 4M categories - (RLI1M, RLI2M, RLI3M, RLI4M),
- **Total Risk Indicator** in production systems (**TRI**).

STAGE 5: Implementation of improvement actions to remove or minimize risk occurrence. Based on Pareto principle – 80% of consequences come from 20% of the causes. Therefore, the implementation of appropriate improvement actions for the upper 20% of risk level indicators should minimize the risk of system disruptions.

4 Risk Assessment for Production Processes – Case Study from Automotive Sector

The case study was conducted in the second half of 2017 in an automotive industry business. It focused on risk management to guarantee the efficiency of production process by losses reduction. The study consists of three months of analysis (June, October, November), when production losses were dominant – Table 2. The assembly line – A1 chosen for the study produces safety valves for truck semi-trailers.

Table 2. Period of production losses's analysis.

Production line A1	6/2017	7/2017	8/2017	9/2017	10/2017	11/2017	12/2017	Sum
Production goal [pcs]	5421	4494	3447	5563	4497	4970	3527	31919
Production output [pcs]	4066	4676	4082	5237	2719	3645	3363	27788
Difference [pcs]	1355	−182	−635	326	1778	1325	164	4131

The goals of the presented analysis were to assess process risks by identifying and categorizing risk factors occurring during production process and to establish corrective activities to improve process efficiency.

In order to specify the reasons of loss in efficiency, hence identification of risk factors on analysed assembly line, the data from hourly control cards were tracked. Production line workers recorded the number of units produced at each working hour. They then compared the result with the target hourly output. If the goal was not achieved, it was necessary to explain the reason and determine the duration of the production line downtime. Following data allowed to determine the cause of the failure in achieving the target production efficiency Based on this simple tool, risk factors have been established along with their frequency of occurrence and duration. The results of analysis of hourly control cards are shown in Fig. 1.

The presented graph shows that the main risk factors are closely connected with each other. The tester's and tool equipment's breakdowns (every month, the breakdown of screwdriver, table's sensor or lubricator has occurred several times) resulted in production of faulty items and consecutive error analysis, as well as in set up of assembly line. During three analysed months, the noted production losses summed up to 157 h. Additionally, the amount of losses has been supplemented with the production leader's data regarding employee's absence. In such cases, assembly line has not been started and hourly control cards have not been filled. The stoppage of line during the analyzed period summed up to almost 404 h.

In the next step of analysis, the identified risk factors have been divided into 4M categories. In the case of line A1, this categorization looked as follows:

- Machine (1M) – 9% (equipment breakdown, tester breakdown, settings/adjustment of the tester after changeover),

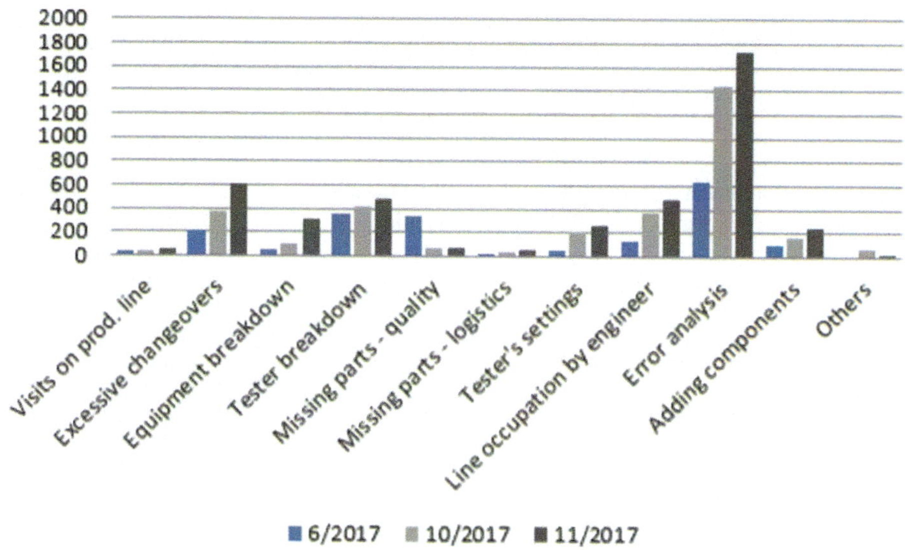

Fig. 1. Production losses of A1 line.

- Material (2M) – 34% (missing components due to logistics or quality issues),
- Method (3M) – 27% (excessive changeovers, production line occupied by engineer, error analysis by assembly operator, adding components, training of employees, visits on production line,)
- Man (4M) – 30% (absence).

Subsequently, the identified risk factors were analyzed according to the frequency of their occurrence, severity of the impact and probability of their detection. The analysis was conducted in accordance with the suggested assessment criteria and its results are presented in Table 3.

Table 3. Analysis of risk factors.

Cat.	No.	Risk factor	Frequency of occurrence		Severity of the effect		Probability of detection	
1M				O1M		S1M		D1M
	$1M_1$	Equipment breakdown - screwdriver	3x6/2017 3x10/2017 3x11/2017	5	Total stoppage: 395 min	4	High probability of risk's factor detection and its reason	3

(*continued*)

Table 3. (*continued*)

Cat.	No.	Risk factor	Frequency of occurrence		Severity of the effect		Probability of detection	
		Equipment breakdown - sensor	1x6/2017 2x10/2017	4	Total break: 50 min	3	Middle high probability of risk's factor detection and its reason	4
		Equipment breakdown - lubricator	1x11/2017	3	23 min	3	Ongoing controls will certainly detect the risk's factor as well as its cause	1
	$1M_2$	Tester breakdown	2x6/2017 1x10/2017 2x11/2017	5	Assembly stoppage for 1258 min in total	6	Medium probability of risk's factor detection and its reason	5
	$1M_3$	Settings of tester after changeover	6x6/2017 16x10/2017 16x11/2017	8	Total stoppage: 517 min	4	Ongoing controls will certainly detect the risk's factor as well as its cause	1
2M				O2M		S2M		D2M
	$2M_1$	Missing parts due to logistic issues	1x6/2017 3x10/2017 3x11/2017	3	Stoppage of 115 min in total	3	Risk factor will certainly be detected	2
	$2M_2$	Missing parts due to quality issues	17x6/2017 4x10/2017 3x11/2017	4	Break of 464 min	4	Risk factor will certainly be detected	2

(*continued*)

Table 3. (*continued*)

Cat.	No.	Risk factor	Frequency of occurrence		Severity of the effect		Probability of detection	
	$2M_3$	Missing parts from suppliers	3x6/2017 1x11/2017	4	Total stoppage 17 shifts	10	V. high probability of detection	1
3M				O3M		S3M		D3M
	$3M_1$	Excessive changeovers	23x6/2017 25x10/2017 33x11/2017	7	In total 1176 min of offline	4	Ongoing controls will certainly detect the risk's factor as well as its cause	1
	$3M_2$	Assembly line occupation by engineer	7x6/2017 13x10/2017 7x11/2017	6	In total 973 min of line stoppage	4	Medium probability of detection via current controls. The detection of the cause of risk is likely	5
	$3M_3$	Error analysis by assembly operator	23x6/2017 25x10/2017 33x11/2017	7	Total stoppage 2507 min	5	Medium probability of detection via ongoing controls. The detection of the cause of risk is likely	5
	$3M_4$	Adding components	8x6/2017 7x10/2017 7x11/2017	6	Total line stoppage 491 min	4	V. high probability of detection	2

(*continued*)

Table 3. (*continued*)

Cat.	No.	Risk factor	Frequency of occurrence		Severity of the effect		Probability of detection	
	$3M_5$	Visits on assembly line	3x6/2017 3x10/2017 4x11/2017	4	Assembly offline 120 min in total	3	Probability of risk factor detection and its cause is v high	2
4M				O4M		S4M		D4M
	$4M_1$	Absence of employees	3x6/2017 1x10/2017 13x11/2017	8	Lack of staff for 16 shifts	10	Probability detection is v low. Minimal supervision of HR	7

The next step was to assess the risk of the analyzed process based on the established formulas [14]. Firstly, value of Single Risk Indicators (SRI) for each category were calculated. Value of SRI for first factor in the Machine category ($1M_1$) - that is equipment breakdown – equals:

$$SRI1M_1 = \sum_{i=1}^{n} (O1M_1 \cdot S1M_1 \cdot D1M_1) \tag{1}$$

hence

$$SRI1M_1 = (5 \cdot 4 \cdot 3) + (4 \cdot 3 \cdot 4) + (3 \cdot 3 \cdot 1) = 117 \tag{2}$$

Value of $SRI1M_1$ for equipment breakdown equals 117 units. Values of SRI for factors second and third correspondingly equal:

$$SRI1M_2 = 5 \cdot 6 \cdot 5 = 150 \tag{3}$$

$$SRI1M_3 = 8 \cdot 4 \cdot 1 = 32 \tag{4}$$

Then, having the values of all risk factors from each category, value of Risk Level Indicator (RLI) for the machine category RLI1M could be calculated:

$$RLI1M = \sum_{i=1}^{n} RLI1M_i \tag{5}$$

hence

$$RLI1M = 117 + 150 + 32 = 299 \tag{6}$$

According to the established formula, value of Risk Level Indicator for the machine category RLI1M is 299 units.

The calculations have been made for remaining categories similarly. The summary of calculations is presented in Table 4.

Table 4. Assessment of risk level.

		O1M	S1M	D1M	SRI1M	RLI1M	TRI
1M	$1M_1$	5	4	3	117	299	1344
		4	3	4			
		3	3	1			
	$1M_2$	5	6	5	150		
	$1M_3$	8	4	1	32		
		O2M	S2M	D2M	SRI2M	RLI2M	
2M	$2M_1$	3	3	2	18	90	
	$2M_2$	4	4	2	32		
	$2M_3$	4	10	1	40		
		O3M	S3M	D3M	SRI3M	RLI3M	
3M	$3M_1$	7	4	1	28	395	
	$3M_2$	6	4	5	120		
	$3M_3$	7	5	5	175		
	$3M_4$	6	4	2	48		
	$3M_5$	4	3	2	24		
		O4M	S4M	D4M	SRI4M	RLI4M	
4M	$4M_1$	8	10	7	560	560	

Value of Total Risk Indicator (TRI) in presented production system is 1344 units. Factor from men category of 4M ($4M_1$- Absence of employees) has the highest input on the total risk value. The factors from the 3M category were also crucial. The smallest share relates to the risk factors for material category – 7%. The last stage was the introduction of improvement actions in order to eliminate or minimize risk occurrence. According to Pareto principle, the threshold of 80% was established. The factors that required immediate improvement were related to employees' absences ($4M_3$), errors analysis by assembly operator, ($3M_3$) tester's breakdown ($1M_2$) and assembly line's occupation by engineer ($3M_2$).

Based on presented risk assessment's results, the analyzed organization implemented actions below:

- human resources monitoring system to track employee holidays etc.
- attendance premium

- conducting training for assembly operators. The training focused on the process of analysing errors for each product and the attendees of this training have been documented for further verification
- introduction of work's instruction related to the analysis of errors for various cases
- performing the analysis of errors 'outside' of assembly line by worker in order to minimize losses
- introduction of a rule that an engineer is not allowed to take the assembly line which has production backlogs. Engineers can perform tests or analysis only during production break.

5 Conclusion

Risk management to improve efficiency of processes, is fairly difficult task. Many enterprises use independently developed methods for calculating the risk. However, in many production systems, procedure of risk management is improper or disregarded at all. This article presents the issue of risk management in production systems. It has been done through providing the new, integrated methodology of risk assessment and analysis with regards to the main 4M categories of factors category disturbing the proper production process.

In order to know which area requires improving the most, it is crucial to identify, categorize and establish the probability of occurrence, the severity and detection rate of each risk. In this way, appropriate actions can be applied to certain areas as an attempt to improve them.

Process of risk management should be a standard routine of any organization. Implementation of one-time analysis will not bring any positive effects. The levels of risk are not constant and may change over time. The risk factors decrease, disappear or new threats appear. However, effective risk management leads companies to improve their activities and add value to the organization [14].

References

1. Owsian, P.: Risk and its impact on the financing of enterprises. Acta Univ. Nicolai Copernici. Econ. **1**, 97–121 (2015)
2. Burduk, A., Musiał, K., Górnicka, D., Kochańska, J.: Production effectiveness improvement with the use of Tabu search. In: CISIM 2019, LNCS 11703, pp. 293–302 (2019)
3. Pająk, E., Klimkiewicz, M., Kosieradzka, A.: Managing Production and Services. PWE, Warszawa (2014)
4. Horváthová, J., Mokrišová, M., Vrábliková, M.: Benchmarking – a way of finding risk factors in business performance. J. Risk Finan. Manag. **14**(5), 221–234 (2021)
5. Bizoń-Górecka, J.: Reliability and Risk Engineering in Enterprise Management. Publishing House of the Center for Organizational Progress, Bydgoszcz (2001)
6. Spath, P.L.: Error Reduction in Health Care: A Systems Approach to Improving Patient Safety, 2nd edn. John Wiley & Sons Inc, San Francisco (2011)
7. Chiarini, A.: Lean Organization: From the Tools of the Toyota Production System to Lean Office, 1st edn. Springer-Verlag Mailand, Milan (2013)
8. PN-ISO 31000:2018–08 Risk Management – guidelines

9. Hopkin, P.: Fundamentals of Risk Management: Understanding, Evaluating and Implementing Effective Risk Management, 5th edn. KPL, London (2018)
10. Kulinska, E.: Selected tools for risk analysis in logistics process. Arch. Transport **24**(1), 27–41 (2012)
11. Gaschi-Uciecha, A.: The essence of risk in logistic processes. Sci. Papers Silesian Univ. Technol. Organ. Manag. **70**, 119–129 (2014)
12. Berisha, Q.A., Kutllovci, E., ShirokaPula, J.: Strategic management tools and techniques: a comparative analysis of empirical studies. Croatian Econ. Surv. **19**(1), 67–99 (2017)
13. EN 31010:2010 Risk management. Risk assessment techniques
14. Rysińska-Wojtasik, D., Burduk, A.: Integrated risk management in production systems. In: Saeed, K., Homenda, W. (eds.) CISIM 2018. LNCS, vol. 11127, pp. 220–229. Springer, Cham (2018). https://doi.org/10.1007/978-3-319-99954-8_19
15. Bagdanavičiūtė, I., Kelpšaitė-Rimkienė, L., Galinienė, J., Soomere, T.: Index based multi-criteria approach to coastal risk assessment. J. Coast. Conserv. **23**(4), 785–800 (2019)
16. Diao, H., Ghorbani, M.: Production risk caused by human factors: a multiple case study of thermal power plants. Front. Bus. Res. China **12**(1), 1–27 (2018). https://doi.org/10.1186/s11782-018-0035-9
17. Chlebus, M., Werbińska-Wojciechowska, S.: Assessment methods of production processes reliability – state of the art. J. KNBiN **41**, 247–276 (2017)
18. Jankajova, E., Kotus, M., Holota, T., Zach, M.: Risk assessment of handling loads in production process. ACTA Univ. Agric. Silvic. Mendel. Brun. **64**, 449–453 (2016)
19. Seitia, H., Fathib, M., Hafezalkotobc, A., Herrera-Viedmade, E., Hameedf, I.A.: Developing the modified R-numbers for risk-based fuzzy information fusion and its application to failure modes, effects, and system resilience analysis (FMESRA). ISA Trans. **113**, 9–27 (2021)
20. Hrbackova, L.: Risk-based thinking in the production process using the methods of quality assurance matrix and the FMEA process. J. Syst. Integr. **1**, 21–28 (2016)
21. Kuroda, T.: A model of stratified production process and spatial risk. Netw. Spatial Econ. Springer **15**(2), 271–292 (2015)
22. Burduk, A.: Assessment of risk in a production system with the use of the FMEA analysis and linguistic variables. In: Corchado, E., Snášel, V., Abraham, A., Woźniak, M., Graña, M., Cho, S.-B. (eds.) HAIS 2012. LNCS (LNAI), vol. 7209, pp. 250–258. Springer, Heidelberg (2012). https://doi.org/10.1007/978-3-642-28931-6_24
23. Zymonik, Z., Hamrol, A., Grudowski, P.: Quality and Safety Management. 1st edn. PWE, Warszawa (2013)
24. Liker, J.K., Meier, D.P.: The Toyota Way. Fieldbook. A Practical Guide to Implementing Toyota's 4P. MT Biznes, Warszawa (2011)
25. Rysińska-Wojtasik, D.: Methodology of risk management in production systems focused on reducing losses, vol. 7, p.169. Reports of the Faculty of Mechanical Engineering of the Wrocław University of Technology (2019)

The Use of Quality Tools to Improve the Risk Management Cycle in the Shaping of the Work Environment

Adam Górny(✉)

Poznan University of Technology, Institute of Safety and Quality Engineering, ul. Prof. Rychlewskiego 2, 60-965 Poznań, Poland
adam.gorny@put.poznan.pl

Abstract. For any efforts to improve working conditions that are guided by risk assessment to be effective, it is critical to rely on accurate information and ensure that any improvement measures taken can be assessed and modified at any stage. Success in achieving the desired effects depends on the use of appropriate tools to help the organization improve its ability to achieve intended outcomes. The paper indicates that is particularly vital to deploy proper instruments to assess, select, and supervise activities so that the work environment to be managed in an orderly manner on the basis of risk management cycle guidelines. The corrective measures will then contribute to the achievement of the intended improvement outcomes in the work environment and possibly provide the organization with competitive advantages. The article discusses based on literature sources the use of tools traditionally associated with quality engineering. Based on relevant literature and field observations in manufacturing organizations, it assesses the option of using quality tools at each stage of risk management. The article focuses on the possibility of achieving outcomes consistent with a particular stage of risk management. Many of the tools discussed in the article can be used to improve the measures attributed to various stages of management, enabling organizations to garner benefits that are characteristic of that management stage rather than the type of tool used. The study refers to the possibility of using the tools for specific activities related to the shaping of the work environment.

Keywords: Improvement · Work environment · Safety · Elimination of hazards and nuisances · Risk management

1 Introduction

Any risks occurring in a work environment, the majority of which are inextricably linked with an organization's processes, require solutions that will mitigate their impact on workers and the severity of the damage they may cause. Such risks may ultimately reduce worker efficiency. Any improvement measures should be designed accordingly to the nature of both the issue at hand and the resulting non-compliance and reflect the magnitude of potential consequences [1–3]. The improvement measures undertaken in response to an issue should be adequate for the nature of the problem, as identified, and ensure effective improvement. This requires that proper account be taken of all

A. Hamrol et al. (Eds.): MANUFACTURING 2022, LNME, pp. 12–22, 2022.
https://doi.org/10.1007/978-3-031-00218-2_2

factors that affect the manufacturing process. Such factors include conditions related to workplace equipment and the individual predispositions of workers assigned to perform work. Moreover, due to an increasing complexity of processes, organizations need to constantly search for solutions to ensure continuous process improvements [4].

The study provides a broad view of the analyzed problem. In particular, close oversight and efforts to identify the need for improvement measures are required where the performance of work may generate high risks caused by particularly hazardous conditions, poor work organization and the use of equipment poorly adapted to needs. The nature of particular issues needs to be recognized to develop responses that will effectively eliminate risks and ensure the safe performance of work. Worker safety during the operation of processes is central to any successful improvement [5, 6]. Improvement measures should not only be guided by risk assessment, but also ensure the effective use of systemic activities included in the occupational risk management loop. Thereby, improving the management process.

Responses to risks may be guided by the ISO 31000 standard [7], which is designed to increase the effectiveness of improvement measures. This is particularly helpful where efforts to improve working conditions are expected to additionally help reduce the costs of risk mitigation [8–10]. An added benefit is a favorable perception of improvement measures by the organization's workers and management. Contributing to the reduction of burdens, including financial burdens related to the failure to ensure working conditions, that provide the efficient implementation of tasks and effectively limiting the potential burdens [11–13].

Research to date is based mainly on the use of single tools to analyze specific problems. It is also possible to consider combining tools in order to obtain optimal conclusions [7].

2 Nature and Characteristics of Risk-Management-Guided Improvement Measures

The main purpose of improvement measures is to effectively protect any individuals who may be exposed to hazards. A possible secondary aim may be to reduce the costs of having a work environment that is poorly adapted to the needs of the workers and any other persons entering production floor. Regardless of the reason why they are taken, improvement measures must reflect the nature of any non-conformities [2, 14].

Risk management guidelines allow one to effectively implement well-justified improvement measures. Such guidelines establish new standards in the implementation of improvement measures, helping organizations to de-ploy them in an orderly and duplicable fashion, ultimately contributing to their greater effectiveness and superior overall outcomes [4]. Such outcomes may include reductions in harm to humans, i.e. any direct adverse impacts of hazards on the health and life of workers [15]. They can also easily be measured against relevant financial costs. This is a consequence of determining the level of burdens, which transfer "human consequences" into financial losses.

The use of measures based on occupational risk assessment allows organizations to respond exclusively to the status quo found at the time of assessment and the circumstances that may arise in the near foreseeable future. To successfully achieve long-term improvements, organizations must identify their issues precisely and roll out solutions and measures that will guarantee stable improvements and offer flexibility to accommodate future changes. This makes the deployment of improvement measures a dynamic activity that involves constant responses to changing conditions [14, 16]. One may therefore view it as a process focused on the continuous improvement of the solutions in place that can be readily equated with the PDCA cycle known in quality engineering. A proper sequence of measures that duly account for improvements of a specific nature will support ongoing modifications accordingly to changes in the environment. Obtaining information about the current state of the work environment should be treated as a factor significantly contributing to ensuring the conditions for the effective performance of professional tasks [17]. The inclusion of the human aspect allows for an individual look at the analyzed issues [15].

To effectively manage occupational risks, organizations only need to recognize the "initial state", as found at the time they first assess their work environment, and ensure the option to perfect their original solutions. Their responses will be evaluated on the basis of the effects of their improvements. Indirectly described by reference to the improvement effects achieved.

The factors affecting an organization's ability to deploy the measures included in its risk management loop are shown in Fig. 1.

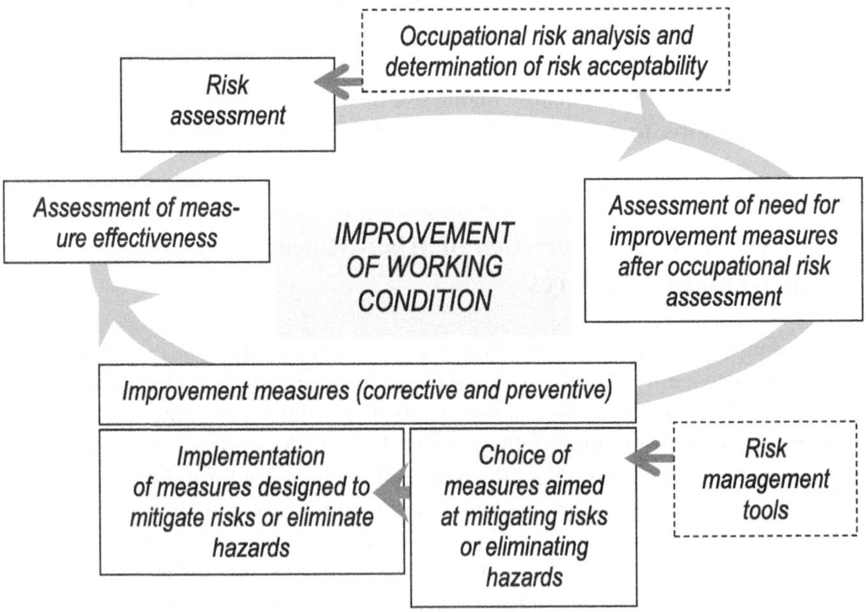

Fig. 1. Occupational risk management loop. Source: Own work.

The response included in the management loop begins with risk assessment. Its purpose is to assess the existing state, or the so-called "starting point", and support decisions that may affect enterprise performance in the long term. Given the considerable uncertainty as to potential future events and their consequences, inaccurate assumptions may be reached leading to misguided choices of measures intended to reduce future risks to an acceptable level. Response effectiveness will therefore largely depend on ongoing monitoring of task performance aimed at bringing risks down to an acceptable degree well into the future [14, 18, 19].

Any improvement measures adopted in line with the risk management methodology to lastingly shape working conditions should recognize key aspects of continuous improvement. These include the requirement to [19, 20].

- Lay the groundwork for management by way of policy formulation and goal setting,
- Select and adopt organizational arrangements that include activity planning and account for mutual relations, responsibilities and rules necessary to roll out solutions as adopted,
- Manage risks in accordance with the adopted action plan that lays down the preconditions to be met to ensure effectiveness,
- Embed risk management procedures in the strategic and operational policies and practices of the organization.

The improvement measures that follow the risk management methodology need to conform to all the above guidelines and be reassessed at every stage of deployment [18]. Any failures to apply such principles may prevent the achievement of the intended outcomes and consequently impede the elimination of non-compliances.

3 Occupational Risk Management

3.1 The Nature of Effective Occupational Risk Management

Risk management combines management and monitoring functions. In practice, risk management should be ensuring the effectiveness of solutions applied in order to eliminate hazards or reduce the consequences of their impact. The main goal of the used activities is to ensure the safety of employees and preventing hazards that may occur in the business, related to the performed tasks affecting the safety and health of the population living in the vicinity of the plant. This applies to the internal and external context of the organization's operation. One should also not forget about the prevention of hazards to the natural environment.

Any improvement measures taken in an environment ridden with hazards and nuisances should be guided by tried-and-tested practices that are known to be both feasible and effective. For the response to existing risks and nuisances to be effective, it is crucial to adjust it for the nature and severity of the non-compliances that evoked that response in the first place, and the feasibility of its deployment in the specific work and business environment [4, 19]. Such a response should be seen as an attempt to reduce accident rates and the incidence of work-related diseases as well as improve working conditions, thus contributing to a reduction in burdens, particularly those resulting from failures to ensure the safe performance of work.

Risk management helps one to make effective decisions in conditions of uncertainty [21]. Any improvement measures based on risk management guidelines require the consideration of prerequisites for effective management. These include the efficiency and effectiveness of the measures taken [21], which is achieved by the proper use of resources, including information [22]. To accomplish the intended outcomes of systemic occupational risk management, organizations should [5, 23]:

- Put in place and maintain hazard identification and risk assessment procedures,
- Regularly assess the methodologies they use and their risk assessment findings,
- Get the workers involved in evaluation activities and, in particular, con-sider their opinions on the nature of non-conformities and on the need for and the required scope of improvement measures,
- Take measures as planned to ensure the suggested solutions are deployed consistently with stakeholder expectations.

Effective risk management measures must extend to all areas of an organization's operations, including its [21]:

- Internal environment, in which the organization pursues its goals,
- External environment, within which the organization operates.

Both the external and internal factors should also be considered when determining the extent of risks and identifying risk features (risk priority determinants) in view of the existing risk policy [21, 24]. Accordingly to the complexity of measures, adequate tools should be selected that best facilitate their deployment and are most likely to secure the achievement of intended out-comes [7].

All such measures are to ensure that the adopted risk management methodology achieves the intended improvement effects.

3.2 Use of Tools to Improve the Efficiency of Improvement Measures

Choosing the right improvement actions is a complex and difficult activity, especially in organizations that do not have sufficient facilities to carry out the indicated actions [25]. Obtaining the intended effects of the change requires the use of tools, measures and methods of improvement appropriate for the area of the working conditions shaping. Whether corrective or preventive, improvement measures must be aligned with the profile of non-conformities and be applied adequately to the severity of risk and the impacts of nuisances. It is vital to consider the circumstances in which a non-compliance occurs and the actual impact they have on the workers.

Conducting improvement actions taken solely on the basis of a description of the current condition is not always sufficient to effectively improve working conditions. [26]. In addition, it is necessary to take into account factors related to human characteristics [27, 28].

When undertaking improvement measures, one must assess the likelihood of achieving the intended outcomes, and therefore use a procedure that will allow one to use the available tools to identify and describe non-conformities and plot a path towards intended outcomes. Such tools should assess the actual need for any selected improvement measures, support their correct choice and application, define how the improvement measure should be used and verify the end results of improvement [23, 29].

The selection of appropriate tools can be based on the normative guidelines indicated in EN IEC 31010 [18] and ISO 31000 standards [21]. But the ultimate criterion for selecting them should be the need for get a specific information.

The quality engineering tools that can be used for this purpose are described in Table 1.

Table 1. Quality engineering tools used in risk management.

Stage of risk management	Sample quality engineering tools	Description of outcomes appropriate for specific risk management stage
Assessment of the need to use for improvement measures	(1) Histogram, (2) Fishbone Diagram, (3) Check sheet, (4) Pareto Diagram, (5) Affinity Diagram, (6) Fault Tree Analysis, (7) Requirements table, (8) Stratification (9) Suzuki Method (10) FMEA	The tools make it possible to: - identify the causes of non-conformities and indicate appropriate preventive measures, - identify areas that require immediate action and improvement, - identify detailed requirements, - collect the available data and organize it to obtain necessary information
Selection of measures that mitigate or eliminate risks	(1) Pareto Diagram, (2) Fishbone Diagram, (3) Matrix Data Analysis, (4) Affinity Diagram, (5) Interrelationship Diagram, (6) Process Decision Program Chart, (7) Suzuki Method, (8) Prioritization Matrix, (9) Scatter Diagram, (10) Relationship Diagram, (11) "Why-Why" Diagram	The tools make it possible to: - determine the need for specific measures, - indicate the best course of action to eliminate a non-compliance, - design a response to multiple non-conformities, - choose the right improvement measures, - guide corrective and preventive measures, - identify the achievable benefits
Deployment of measures that eliminate or mitigate risk	(1) Flowchart, (2) Arrow Diagram, (3) Process Map, (4) Top-Down Flowchart, (5) Deployment Flowchart (6) Pareto Diagram	The tools make it possible to: - identify mutual links between the measures taken, - present details of measures being taken and thereby facilitate their rollout, - indicate the structure of tasks performed and mutual interactions among their components, - make the perception of activities more accurate

(*continued*)

Table 1. *(continued)*

Stage of risk management	Sample quality engineering tools	Description of outcomes appropriate for specific risk management stage
Assessment of effectiveness of undertaken improvement measures	(1) Value Added Analysis, (2) Histogram, (3) Control Chart, (4) Customer Satisfaction Index Method	The tools make it possible to: - evaluate the effectiveness of measures and identify options for their improvement, - identify the need to improve the measures being taken, - assess worker satisfaction with the changes made

Source: Own work based on [1, 4, 14]

4 Discussion

4.1 The Benefits of Using Tools Designed to Improve the Efficiency of Occupational Risk Management

The main purpose of introducing risk management and the use of tools to improve risk management is to obtain benefits derived from the applied improvement actions. Identified risk factors define the need and define the scope of activities [3]. The potential benefits can be measured based on the increased capacity to effectively eliminate such irregularities in the work environment as arise in the performance of work [4, 30]. Such an increase is achieved as a consequence of risk assessment and of a traditional rollout of improvement measures that does not rely on the continuous improvement cycle. Ensuring the possibility of verifying the activities as part of risk management requires the use of objective criteria for assessing the achieved effects.

By managing risks and doing so to an ever higher quality standard, an organization stands to secure benefits that are tantamount to the effects of selecting and employing quality tools that are adequate to a given stage of risk management (see: Table 1, Column 3). Influencing the effectiveness of the improvement actions. Such benefits include [5, 29]:

- Improved chances of achieving desired improvement objectives,
- Support for proactive business management that is responsive to changes in the work environment,
- Raised worker awareness of the need for proper risk handling,
- Improved identification of irregularities and risks and improved opportunities with due account taken of all existing conditions (internal and external),
- Improved principles to govern the performance of manufacturing and ser-vice activities and the resulting improvements in such performance,
- Greater stakeholder confidence in the organization, which is recognized as caring about its workers, its internal work environment and the external environment in which it operates,

- An improved system for the monitoring of impacts on the environment and the workers,
- Improved measures taken to reduce potential losses,
- Better performance in occupational safety, including lower costs of accidents.

Should be considered that the necessary condition for effective occupational risk management is the use of information from verified sources, observations, prognosis and opinions. When obtaining information, one should take into account the feedback that indicates the possibility of discrepancies affecting the perception of risk and estimating its level. It is imperative that the obtained information describes the current state in the most complete and adequate manner. Effective management of occupational risk requires [6, 16–18, 31]:

- Conducting activities related to shaping working conditions in accordance with the safety policy adopted in the organization, defined by the company's top management,
- Basing the improvement activities carried out on the exhaustive analysis of the causes of accidents, near-accidents and work-related diseases,
- Identification of the requirements defining the scope and nature of the necessary improvement actions, allowing for compliance with the applicable legal and normative regulations,
- Planning activities to eliminate or reduce the occurrence of hazards and their implementation in a manner consistent with the principles adopted in the company, based on the guidelines of the safety policy,
- Ensuring monitoring of the improvement activities, aimed at verifying the effectiveness and obtaining the possibility of introducing changes and modifications contributing to the improvement of the possibility of obtaining the desired effects,
- Conducting reviews of the improvement actions to indicate the directions of necessary changes, consistent with the needs resulting from the experience acquired during their implementation.

By deploying risk mitigation measures comprehensively, an organization is enabled to develop a consistent risk mitigation policy. The available scientific studies indicate many of the effects obtained, which primarily relate to the effectiveness of the used tools. However, every action that reduces a risk comes with a financial cost. Risk management allows one to apply an evaluation algorithm that is helpful in identifying and applying the most effective solutions [32]. In this sense, integrating occupational risk management into existing processes will improve their implementation and help to increase long-term benefits, thus providing the company with a competitive advantage [33]. This approach may vary depending on the model and tool employed to assess both the circumstances at hand and the workers' capacity to perform their tasks. In many cases, it requires a variety of tools, not just one. The implication of several tools allows for a broader look at the analyzed problem, obtaining objective comments.

5 Conclusion

Occupational risk is an integral part of every professional activity. It cannot be ignored unless the solutions in place already ensure full protection of workers. In an effort to find a way to mitigate the negative impact of risks on workers and increase the effectiveness of improvement measures, an organization may follow management guidelines that will allow it to modify working conditions by focusing on the management factors that are designed to in-crease management effectiveness.

Risk management should be viewed as a way to ensure occupational safety by improving the effectiveness and efficiency of risk elimination or mitigation with a view to improving working conditions. By incorporating occupational risk management into existing processes, one will improve their performance and help increase long-term benefits, thus providing the company with competitive advantages.

Risk management should be recognized as part and parcel of organization-al improvement whose outcomes depend on the effective performance of critical measures. The success of this process requires ongoing adaptations to changing conditions. The use of tools to identify the problem, indicate improvement actions that are reasonable to take, and evaluate the results obtained. This is the way to achieve this goal.

Acknowledgement. This paper has been prepared as a part of Research Topic No. 0811/SBAD/1051 entitled: "Research on the conditions of the functioning and adaptation a social engineering systems to the implementation of challenges for sustainable development (SD) and Industry 4.0 (I 4.0)".

References

1. Russ, K.: Risk assessment in the uk health and safety system: theory and practice. Saf. Health Work **1**(1), 11–18 (2010)
2. Blanc, F., Escobar Pereira, M.M.: Risks, circumstances and regulation – historical development, diversity of structures and practices in occupational safety and health inspections. Saf. Sci. **130**, 104850 (2020)
3. Trillo-Cabello, A.F., Carrillo-Castrillo, J.A., Rubio-Romero, J.C.: Perception of risk in construction. Exploring the factors that influence experts in occupational health and safety. Saf. Sci. **133**, 104990 (2021)
4. Stürmlinger, T., Jost, D., Mandel, C., Behrendt, M., Albers, A.: Impact and risk analysis in the integrated development of product and production system. Proc. CIRP **91**, 627–633 (2020)
5. Górny, A.: Use of quality management principles in the shaping of work environment. In: Stephanidis, C. (ed.) HCI 2015. CCIS, vol. 529, pp. 136–142. Springer, Cham (2015). https://doi.org/10.1007/978-3-319-21383-5_23
6. Grznár, P., et al.: Modeling and simulation of processes in a factory of the future. Appl. Sci. **10**(13), 4503 (2020)
7. Górny, A.: The use of quality tools in risk management. In: Arezes, P., et al. (eds.) Occupational Safety and Hygiene, SHO 2017, pp. 98–99. Portuguese Society of Occupational Safety and Hygiene (SPOSHO), Guimarães (2017)
8. Sousa, V., De Almeida, N.M., Dias, L.A.: Risk-based management of occupational safety and health in the construction industry. Part 1: background knowledge. Saf. Sci. **66**, 75–86 (2014)

9. Krynke, M.: Risk management in the process of personnel allocation to jobs. In: Ulewicz, R., Nikolić, R.R. (Eds.), System Safety: Human – Technical Facility – Environment, pp. 82–90, Sciendo, De Gruyter, Berlin (2020)
10. Olechowski, A., Oehmen, J., Seering, W., Ben-Daya, M.: The professionalization of risk management: What role can the ISO 31000 risk management principles play? Int. J. Project Manage. **34**(8), 1568–1578 (2016)
11. Shalin, R.T.: Economic cost of occupational accidents: evidence from a small island economy. Saf. Sci. **47**(7), 973–979 (2009)
12. Salguero-Caparrós, F., Pardo-Ferreira, M.C., Martínez-Rojas, M., Rubio-Romero, J.C.: Management of legal compliance in occupational health and safety. A literature review. Saf. Sci. **121**, 111–118 (2020)
13. Restrepo, C.E., Simonoff, J.S., Zimmerman, R.T.: Causes, cost consequences, and risk implications of accidents in US hazardous liquid pipeline infrastructure. Int. J. Crit. Infrastruct. Prot. **2**(1–2), 38–50 (2009)
14. Berner, C.L., Flage, R.: Creating risk management strategies based on uncertain assumptions and aspects from assumption-based planning. Reliab. Eng. Syst. Saf. **167**, 10–19 (2017)
15. Lawrie, M., Parker, D., Hudson, P.: Investigating employee perceptions of a framework of safety culture maturity. Saf. Sci. **44**(3), 259–276 (2006)
16. Probst, T.M., Graso, M., Estrada, A.X., Estrada, G.S.: Consideration of future safety consequences: a new predictor of employee safety. Accid. Anal. Prev. **55**, 124–134 (2013)
17. Morillas, R.M., Rubio-Romero, J.C., Fuertes, A.: A comparative analysis of occupational health and safety risk prevention practices in Sweden and Spain. J. Safety Res. **47**, 57–65 (2013)
18. EN IEC31010:2019: Risk management. Risk assessment techniques, CENELEC, Brussels (2019)
19. Uhrenholdt-Madsen, C., Kirkegaard, M.L., Dyreborg, J., Hasle, P.: Making occupational health and safety management systems 'work': A realist review of the OHSAS 18001 standard. Saf. Scie. **129**, 104843 (2020)
20. ISO Guide 31:2009, Risk management. Vocabulary, ISO, Geneva (2009)
21. ISO 31000:2018, Risk management. Guidelines, ISO, Geneva (2018)
22. Wang, B., Wang, Y.: Big data in safety management: an overview. Saf. Sci. **143**, 1054–14 (2021)
23. Górny, A.: Guidelines for the use of quality engineering tools to improve occupational safety. the case of polish construction companies. In: Arezes, P.M., et al. (eds.) Occupational and Environmental Safety and Health II. SSDC, vol. 277, pp. 31–39. Springer, Cham (2020). https://doi.org/10.1007/978-3-030-41486-3_4
24. Tchiehe, D.N., Gauthier, F.: Classification of risk acceptability and risk tolerability factors in occupational health and safety. Saf. Sci. **92**, 138–147 (2017)
25. Itani, T.: Participatory work improvement in small workplaces. Ind. Health **49**(4), 401–402 (2011)
26. Dong, Y., Hu, S., Zhu, J.: From source credibility to risk perception: how and when climate information matters to action. Resour. Conserv. Recycl. **136**, 410–417 (2018)
27. Kariuki, S.G., Lowe, K.: Integrating human factors into process hazard analysis. Reliab. Eng. Syst. Saf. **92**(12), 1764–1773 (2007)
28. Górny, A.: Man as internal customer for working environment improvements. Proc. Manuf. **3**, 4700–4707 (2015)
29. Gołaś, H., Mazur, A., Mrugalska, B.: Application of risk analysis and quality control methods for improvement of lead molding process. Metalurgija **55**(4), 811–814 (2016)
30. Rajkumar, I., Subash, K., Raj Pradeesh, T., Manikandan, R.M., Ramaganesh, M.: Job safety hazard identification and risk analysis in the foundry division of a gear manufacturing industry. Mater. Today Proc. **46**(17), 7783–7788 (2021)

31. Górny, A.: The use of working environment factors as criteria in assessing the capacity to carry out processes. In: MATEC Web of Conferences (CoSME 2016, the 4th International Conference on Computing and Solutions in Manufacturing Engineering 2016), vol. 94, p. 06011 (2017)
32. Papazoglou, I.A., Aneziris, O.N., Bellamy, L.J., Ale, B.J.M., Oh, J.I.H.: Multi-hazard multi-person quantitative occupational risk model and risk management. Reliab. Eng. Syst. Saf. **167**, 310–326 (2017)
33. Maletič, D., Pačaiová, H., Nagyová, A., Maletič, M.: The link between asset risk man-agement and maintenance performance: a study of industrial manufacturing companies. Qual. Innov. Prosp. **24**(3), 50–69 (2020)

Application of the PFMEA for Risk Assessment in the Production Process

Dariusz Plinta(✉) and Aleksandra Greń

University of Bielsko-Biala, Willowa 2, 43-309 Bielsko-Biała, Poland
dplinta@ath.bielsko.pl

Abstract. Nowadays, risk is inherently related to managing an organization, regardless of its size and business profile. When implementing a strategy of effective enterprise management in a turbulent environment, it is necessary to apply an approach that includes comprehensive risk management in the organization, particularly in the implemented production processes. The latest update of the ISO 9001 and the IATF 16949 standards focused in particular on implementing a risk-based approach by organizations. The ISO 9001 standard guides organizations into applying a multidisciplinary approach to risk management in order to increase their adaptability in relation to changes taking place in both external and internal environment. The article presents an example of a risk management approach in a cast iron foundry based on the use of qualitative tools in order to improve the identification and elimination of both faults occurring in the process and their causes. The PFMEA process analysis is one of the risk management methods in the production process widely used in the automotive industry. The PFMEA analysis is an effective tool for identifying problematic areas, which contributes to ensuring the required quality of foundry products and improving the production process.

Keywords: PFMEA · Quality · IATF 16949 · Casting · Automotive industry

1 Introduction

Risk is an inseparable element of the processes existing in the organization today, as it has a significant impact on the functioning of enterprises. In order to run an organization effectively and efficiently, it is necessary to manage risk at every operational level. By default, a risk management process is aimed at supporting entities in making informed decisions and implementing a given business strategy. By using an interdisciplinary approach to risk management, enterprises increase their adaptability to changes in a turbulent environment. Entities conducting business activity need to understand the importance of targeting their strategy, taking into consideration the mechanisms that enable an early identification of the existing threats and taking adequate corrective and preventive action. In order to ensure stable development in a dynamic environment, enterprises have to take into account the necessity to integrate a risk-based approach into their organizational culture [1].

A. Hamrol et al. (Eds.): MANUFACTURING 2022, LNME, pp. 23–35, 2022.
https://doi.org/10.1007/978-3-031-00218-2_3

According to PN ISO 31000 standard, risk management is a process which systematically applies management policy, procedures and adopted practices for the following activities: setting the context, identification, analysis, evaluation, communication and consultation, proceeding, monitoring and supervision (Fig. 1). At the identification stage, the main challenge for the organization is to obtain as much data as possible about potential risk factors and their consequences for the entity. The result of the identification step should be the compilation of the possibly most complete list of threats resulting from potential events. The aim of the second stage is to analyse the identified risk, which is then subject to evaluation. Risk assessment is the estimation of the degree of threat of individual factors by comparing the established levels of risk with the recognized criteria. The collected data constitute the basis for making decisions related to the methods of dealing with risk.

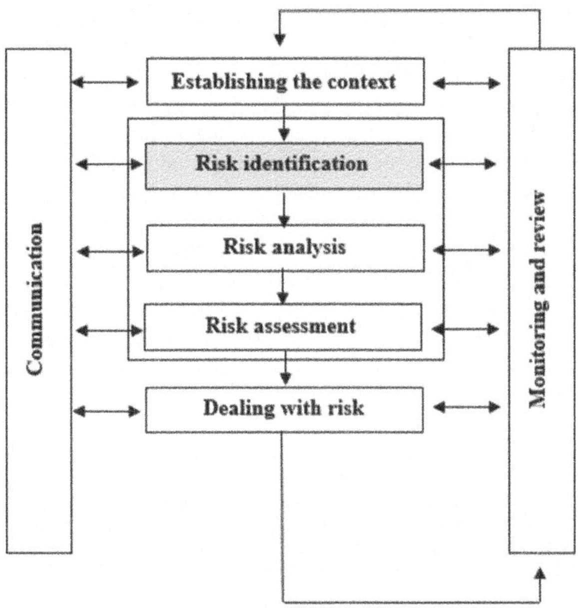

Fig. 1. A risk management scheme according to PN ISO 31000

The purpose of the risk management process is to minimize the impact of threats and to enable the enterprise to take advantage of opportunities. Continuity and order are important characteristics ensuring the effectiveness of the process. It is also important to correlate this process with the organizational culture existing in the company in order to ensure its greatest efficiency. The risk management process should also be carried out in accordance with the adopted methodology and applicable legal regulations, and be understandable to those involved in it [2]. In addition, the importance of focusing on building effective communication between the people involved in the risk management process in the enterprise is definitely worth mentioning. A comprehensive approach to communication during performing risk management activities makes it possible to understand the cause-effect relationships and make the right decisions. The risk management process is primarily aimed at achieving by organizations an acceptable level of risk and creating optimal conditions for further growth of the business entity [1]. The risk-based approach in all sections of the enterprise is a concept that has been strongly popularized since the introduction of the latest amendment to the ISO 9001 and IATF 16949 standards dedicated to the automotive industry. Nowadays, the ISO standards concerning management systems constitute a guideline embedded in the organizational culture of the vast majority of economic entities from the sector of large and medium-sized enterprises. The ISO quality standards are very popular and are the feature the most widespread approach to organization management in the world [3, 4].

The main purpose of applying risk management in the production process is to develop effective and efficient procedures to ensure the continuity of the process and reduce the level of risk to an acceptable level. In order to achieve this goal, it is necessary to acquire the knowledge needed to accurately forecast the effects of potential events in technological processes and to determine the probability of their occurrence based on risk assessment methods.

Thanks to the increasing popularity of this subject, the methods and tools applicable in the risk management process are being constantly improved. Most of the methods dedicated to risk assessment are based on the diagram shown in Fig. 1. Today, enterprises dispose of an extensive range of methods to support the application of a risk-based approach, which are based both on statistics, expert knowledge and experience. Many of them are recommended in international standards, guidelines and recommendations of international associations or regulated by law. One of the preventive quality management methods is the FMEA analysis, which is irreplaceable in the context of meeting the requirements of ISO 9001: 2015 and IATF 16949: 2016, which impose the requirement to conduct risk analyzes, in particular in production processes. The PFMEA analysis presented in this article is a tool supporting the organization in the risk management processes. The article presents a new approach to conducting PFMEA analysis presented in the FMEA manual in 2019, which is currently being implemented in many manufacturing companies [5, 6].

2 The PFMEA Analysis

Eliminating the threats in the process enables to consistently improve products quality by removing the causes of their occurrence. One of the methods for risk analysis is the PFMEA method, which is widely used in the automotive industry. It focuses on preventing the consequences of defects that may occur in the production phase. The essence of the PFMEA is to establish cause-and-effect relationships of potential defects in the process, taking into account risk factors.

Failure cause and effect analysis (FMEA) is an analytical methodology used to ensure that potential failures are considered and resolved throughout the course of product and process development. The application of the FMEA methodology enables to reduce production costs. This is because the cost of removing a non-conformities increases exponentially as the product life cycle progresses. Consequently, the later a non-compliance is discovered, the more expensive it will be to remove it [7, 8].

Until now, AIAG has developed its requirements for the automotive industry in companies cooperating with the American industry, and VDA issued a standard that presented a risk analysis from the point of view of the German automotive industry. It resulted in the necessity to conduct risk analyses according to various approaches, depending on the clients' expectations. In addition to developing a consensus between the various approaches used in the automotive industry, the new standard is also intended to solve the problems affecting the added value that companies have so far struggled with when conducting risk analyses. One of the objections to the latest version of the FMEA methodology was that some of the teams responsible for conducting the analysis did not take into account all the defects and all the effects when running the operations. The causes of the problem were, among others, deliberate actions and the lack of sufficient knowledge to identify the complexity of the analysed problems. The new manual requires a detailed description of the elements that make up the analysed process, starting with the manufactured product, through the phases of the process, and ending with the elements of the process. Problems can be eliminated by using the defect tree tool, which significantly reduces the risk of omitting a defect in the analysed area. Another change aimed at improving the completeness of the analyses carried out is the requirement to provide information about the expected and required functions [7].

Another significant change introduced in the new edition of the FMEA manual is the replacement of the RPN coefficient, previously used to establish priority actions, by the AP matrix. The change was caused by the basic disadvantage of the RPN coefficient, which was flattening the risk image as a result of non-compliance. AP is not a value calculated like the RPN, which was the product of three indices. The AP risk factor is determined from a special table. The new manual systematizes the way the FMEA is carried out by introducing a simple standard called 5T.

The two main types of FMEA are Design FMEA (DFMEA), which focuses on product analysis, and Process FMEA (PFMEA). The AIAG & VDA FMEA manual describes the above approaches, with the difference that it introduces a completely new area of FMEA-MSR risk analysis (FMEA for Monitoring and System Response), which is related to the requirements of ISO 26262. The purpose of this methodology is to estimate defects in electronic and electrical systems and features related to the function of maintaining safety and compliance with legal regulations. An appropriate use of the supplementary FMEA-MSR is also intended to ensure compliance with environmental regulations. The seven-step FMEA-MSR process is very similar to DFMEA, except for the methodology of Step Four (Failure Analysis) and Step Five (Risk Analysis).

The new FMEA according to AIAG & VDA has been transformed into a seven-step process. Most of the steps are already known from the previous FMEA approach. What is new, however, is defining the scope of the FMEA analysis in the first step and documenting the results by the use of lessons learned.

Table 1. The 7-step FMEA approach. Source: Own study

7 STEPS ACCORDING TO FMEA AIAG & VDA						
SYSTEM ANALYSIS			ERROR ANALYSIS AND RISK ASSESSMENT			REPORTING
1st step	2nd step	3rd step	4th step	5th step	6th step	7th step
Planning and definition of scope (5T)	Structure analysis	Function analysis	Error analysis	Risk analysis	Optimization (Risk reduction)	Documenting the results
Planning and preparation. Defining the scope of the analysis.	Product / process structure analysis using a flow chart.	Overview of the functionality of the product / process. Establishing the requirements.	Creating a chain of defects with their causes.	Conducting a risk assessment based on the (S, O, D) coefficient and the AP matrix.	Implementation of corrective and preventive actions. Risk reassessment.	Documenting the results of the analysis. Application of lessons learned

In order to improve the prepared documentation, in the seventh step a report should be prepared, which should contain, first of all, a summary of the findings, an overview of the risk analysis and confirmation of closing the activities. The introduction of the first step aims at:

- essentially defining the scope of the project,
- clarifying the type of the used FMEA,
- determining the time limits of the FMEA analysis, and
- defining the basis for further stages of the process according to 5T, i.e. Team (FMEA team), Timing, inTent (aim of the FMEA), Tools and Tasks.

In the fifth step, risk assessment is carried out by assigning individual values, in accordance with the guidelines in Tables 2, 3 and 4 below, for the severity, probability of a defect occurrence and detectability, after which the risk level is defined in accordance with the AP matrix. The current edition of the FMEA manual introduces a new form template, which is characterized by a highly extensive amount of information that needs to be documented (change from a dozen to almost thirty columns) [8–11].

The FMEA analysis should be performed by an interdisciplinary team under a leader's supervision. The team should include people responsible for the implementation of the project, specialists in a given area (quality, production, technology, supply, etc.). The effectiveness of the analysis is also enhanced by including representatives of customers and suppliers in the work of the team. The FMEA analysis process should be started at least if:

- there has been a change in a product, production process, material flow, assembly method,
- the process does not have the required capacity,
- new devices, tools or machines have been applied,

Table 2. Evaluation criteria for the severity indicator for PFMEA according to the manual FMEA AIAG & VDA ed. 1. Source: FMEA AIAG & VDA manual

Severity	Your Plant Severity criteria	Ship to Plant Severity criteria	Customer Severity criteria	Corporate or product Line Examples
10	Failure may endanger operator (machine or assembly), Possible long-term effects on health of production associates	Failure may endanger operator (machine or assembly), Possible long-term effects on health of production associates	Affects safe operation of the vehicle and/or other vehicle, the heath of operator or passenger(s) or road users of pedestrians.	
9	Failure may result in in-plant regulatory noncompliance	Failure may result in in-plant regulatory noncompliance	Noncompliance with regulations	
8	100% of product affected may have to be scrapped	Line shutdown greater than full production shift. Stop shipment possible. Field repair or replacement required (Assembly to End User) other than for regulatory noncompliance	Loss of essential vehicle function necessary for normal driving during expected service life.	
7	A portion of the production run may have to be scrapped. Deviation from primary process; decreased line speed or added manpower	Line shutdown 1 hour ~ Full Production Shift. Stop shipment possible. Field repair or replacement required (Assembly to End user) other than for regulatory noncompliance	Degradation od essential vehicle function necessary for normal driving during expected service life.	
6	100% of production run may have to be reworked off line and accepted	Line shutdown up to one hour	Loss of convenience function	
5	A portion of the production run may have to be reworked off line and accepted	Less than 100% product affected. Strong possibility for additional defective product-sort required. No Line Shutdown	Degradation of convenience function	
4	100% of production run may have to be reworked in station before it is processed	Defective product triggers significant reaction plan. Additional defective products not likely. Sort not required	Perceived quality of appearance, sound or haptics unacceptable to most customers	
3	A portion of the production run may have to be reworked in station before it is processed	Defective product triggers minor reaction plan. Additional defective products not likely. Sort not required	Perceived quality of appearance, sound or haptics unacceptable to many customers	
2	Slight inconvenience to process, operation, or operator	Defective product triggers no reaction plan. Additional defective products not likely. Sort not required. Requires feedback to supplier	Perceived quality of appearance, sound or haptics unacceptable to some customers	
1	No discernible effects	Defective products not likely. Sort not required. Feedback to supplier not required	No discernible effect	

Table 3. Evaluation criteria for the indicator of probability of defect occurrence for PFMEA according to the manual FMEA AIAG & VDA ed. 1. Source: FMEA AIAG & VDA manual

OCC	Estimated Occurrence	Process Experience	Prevention Controls	Corporate or prod-uct Line Examples
10	Occurrence during manufacturing or assembly cannot be determined, no preventive controls, or occurrence during manufacturing or assembly is extremely high	New process without experience. New product application	Est practices and procedures do not exist	
9	Very high occurrence during manufacturing or assembly	Limited experience with the process. Application significantly different from previous application	Not targeted to specific failure cause. Newly developed for this process. First application of new procedures with no experience	
8	High occurrence during manufacturing or assembly	Known but problematic process. Application presents significant process challenges	Not a reliable prevention of the failure cause. Few existing procedures and best practices, not directly applicable for this process.	
7	Moderately High occurrence during manufacturing or assembly	Similar process with evidence of nonconformance in excess of accepta-	Provides limited use in preventing a failure cause. Procedures and best practices apply to the baseline pro-	
		ble rate. No experience with this application in the company	cess, but not the innovations	
6	Moderate occurrence during manufacturing or assembly	Similar process with some evidence of nonconformance. Limited experience with this application in the company	Provides some ability to prevent a failure cause. Procedures and best practices exist but are insufficient to ensure that the failure will not occur	
5	Moderate occurrence during manufacturing or assembly	Similar process with successfully completed process validation. Limited experience with supplication at this facility	Capable of finding deficiencies in the process. Process design addresses lessons learned from previous designs. Best Practices re-evaluated for this process, but have not yet been proven. Provides some indication that the process will not yet been proven. Provides some indication that the process will not have problems.	
4	Moderately low occurrence during manufacturing or assembly.	New setup based on proven process. Application does not introduce significant risk of process challenges.	Capable of finding deficiencies in the process related to the failure. Predecessor process and changes for new process conforms to best practices and procedures. Indicates likely process conformance.	
3	Low occurrence during manufacturing or assembly	Process has been tried and tested with successful results in series production. History of capability within control limits. Similar application	Capable of finding deficiencies in the process related to the failure. Process expected to confirm to best practices and procedures, considering Lessons Learned from previous processes. Predicts conformance of production design	
2	Very low occurrence during manufacturing or assembly.	Process has been tried and tested with successful results in series production. History of capability within control limits. Carryover application	Capable of finding deficiencies in the process related to the failure. Process expected to conform to best practices, considering Lessons Learned from previous processes, with significant margin of confidence. Indicates confidence in design conformance.	
1	Possibility of failure is eliminated through preventative control and history of failure-free series production. The failure cannot occur in series production	Cause cannot occur because failure is eliminated through demonstrated preventative control	Failure cannot occur in series production. Process proven to conform to procedures and Best Practices, considering Lessons Learned	

Table 4. Evaluation criteria for the detectability indicator for PFMEA according to the manual FMEA AIAG & VDA ed. 1. Source: FMEA AIAG & VDA manual

DET	Ability to Detect	Detection criteria	Corporate or product Line Examples
10	Absolute uncertainty	The failure will not or cannot be detected as no testing or inspection method has been established or is known	
9	Very remote	Failure is not easily detected. Random audits<100% of product. It is unlikely that the testing or inspection method will detect a possible malfunction or fault mechanism.	
8	Remote	Defect (Failure Mode) detection downstream through visual, tactile or audible means. Ability of testing or inspection method is uncertain or the company/business unit has no experience with the defined testing or inspection method. The method relies on a human for verification and disposition	
7	Very Low	Defect (Failure Mode) detection in-station through visual, tactile or audible means. Ability of testing or inspection method is very low or the company/business unit has little experience with the defined testing or inspection method available. The method relies on a human for verification and disposition.	
6	Low	Defect (Failure Mode) detection downstream through use of variable gauging (e.g. calipers, dial gauge, etc.) or attribute gauging (e.g. go/no-go, manual torque check/clicker wrench, etc.). Ability of testing or inspection method not been proven for this application. The company/business unit has experience with the defined testing or inspection method. Test/inspection/measuring equipment capability in not yet proven	
5	Moderate	Defect (Failure Mode) or Error (Failure Cause) detection in station through use of variable gauging (calipers, dial gauge, etc.) or attribute gauging (go/no-go, manual torque check/clicker wrench, etc.). Proven testing or inspection method for comparable products under new operating/boundary conditions. Test/inspection/measuring equipment capability for comparable processes is confirmed through gauge repeatability and reproducibility evaluations. Fer set-up Causes only: confirmation of setup with first piece check and use of last piece check, as applicable	
4	Moderately high	Defect (Failure Mode) detection downstream through use of controls that will detect and control discrepant product. Proven testing or inspection method from comparable processes under similar operating/boundary conditions (machines, material). Test/inspection/measuring equipment capability grom comparable processes confirmed through gauge repeatability and reproducibility evaluations. The required error proofing verification is performed.	
3	High	Defect (Failure Mode)detection in station through use of controls that will detect and control discrepant product. Proven testing or inspection method from comparable processes under similar operating/boundary conditions (machines, material). Test/inspection/measuring equipment capability from comparable processes confirmed through gauge repeatability and reproducibility evaluations. The required error proofing verification is performed.	
2	Very high	Error (Failure Cause) detection in station through use of controls that will detect error and prevent discrepant product from being produced. Proven testing or inspection method from identical processes under the same operating/boundary conditions (machines, material). Test, inspection/measuring equipment capability from identical processes confirmed through gauge repeatability and reproducibility evaluations. The required error proofing verification is performed.	
1	Almost certain	Discrepant product cannot be physically produced due to design (part geometry) or process (fixture or tooling design). The effectiveness was demonstrated on this product.	

- production has been relocated,
- there has been an accident during work,
- a customer has filed a complaint,
- as part of regular maintenance.

Continuous improvement is a necessary condition for companies that want to build a strong position on the market and competitive advantage. The use of the FMEA analysis is aligned with the ideology of continuous improvement, as it enables to eliminate defects occurring in the process consistently and effectively. PFMEA makes it possible to prioritize activities, which is important considering limited resources of the organization. Considering the variability of environmental conditions, improving processes and the preventing faults should be a continuous process aimed at achieving high efficiency and effectiveness. To sum up, by using quality management instruments such as the FMEA methodology, it is possible to effectively detect defects in the production process, minimize production costs of a product and introduce effective corrective actions, which allow, among others, to reduce losses and direct costs in production [12, 13].

3 The PFMEA Analysis of the Production Process

Fierce market competition and increasing customer requirements mean that foundries place great emphasis on the production of castings of high quality, reliability and operational durability. In order to meet these requirements, it is necessary to identify potential defects in the process by conducting risk analyses [14]. Metalpol is a producer supplying products for the railway, automotive, machinery, mining, agricultural, construction and public works sectors, made of grey cast iron EN-GJL-200, 250, 300 and ductile iron EN-GJS-400-15, 500-7, 600-3, 700-2. The plant also manufactures industrial fittings.

Fig. 2. A vertical moulding line

The process of casting iron in bentonite moulds in the enterprise consists in pouring liquid metal into a bentonite mould under the action of gravity (Fig. 2). It is the oldest casting technology. The most important issues considered in this technology include, among others: the forming method, the division of the cores, the analysis of conditions for the castings solidification, the review of the arrangement of the cast in the form including the division of the mould, and the analysis of the supply system, arrangement, dimensions of the elements the infusion system and the range of their impact [14].

There is a wide variety of casting defects that result from the gravity casting technology itself. Typical casting defects occurring with this technology result from the phenomenon of foundry shrinkage. In addition to shrinkage defects, the disadvantages of the discussed technology include non-metallic precipitates such as slagging, gas bubbles and gas bubbles, and sandiness. In the case of castings, where the quality of the product is influenced by a multitude of closely related factors, it is very difficult to determine the root cause of the problem. In order to conduct the PFMEA analysis effectively, it is necessary to be familiar with the basic issues related to the conduct of foundry processes by an interdisciplinary team [15].

The main assumption of this paper is to present the implementation of the risk management process in a manufacturing company through the use of an exemplary risk analysis, which is PFMEA. In the conducted research described in chapter three, in order to prevent potential inconsistencies in the production process, Metalpol has applied the PFMEA methodology. Table 5 shows a fragment of PFMEA for the casting process carried out on a vertical moulding line. The following analysis was carried out based on the guidelines and principles presented in the latest FMEA handbook, issued in June 2019. The FMEA analysis was carried out in accordance with the 7-step process illustrated in Table 1. The new FMEA manual requires from the organization to introduce a new way of conducting FMEA analysis, which may be a problem at the beginning, especially with the precise definition of the structure of analysis and the functions of its individual elements. The individual indicators of severity, occurrence and detection were assessed by an interdisciplinary team on the basis of Tables 2, 3 and 4. This table includes corrective and preventive actions aimed at ensuring the expected course of the production process by eliminating potential threats.

Table 5. PFMEA analysis. Source: Own study

Structure analysis (step 2)			**Function analysis (step 3)**			**Failure analysis (step 4)**		
1. Process element, System, subsystem, component or process name	2. Process step, Job number, name of the central element	3. Element 4M of the process	1 Function of a process element, system, subsystem, component, or process	2. Process step function and product characteristics	3. Function of the working element of the process and characteristics of the process	1. Failure effect (FE) on the next recipient and / or end user	Severity (S) FE	2. The type of error (FM) resulting from the central element
Core production	Coating of the cores (Dipping bath)	Machine: Dipping bath; Material: Cover, Core; Method: Coating method, Work instruction, Identification label; Person: Worker of Core room	Production of cores necessary to recreate the internal shapes of castings	Protection of the cores, Ensuring a lower surface roughness of the castings	Machine: Coating the core with a coating; Material: Adequate core protection; Method: Presenting the Employee with the quality requirements and the method of performing the operation, Identification of materials; Person: Checking the implementation of the operation	Occurrence of a non-compliant product (surface defects)	8	Drip on the surface of the coated core
Forming	Assembling the cores to the mold (Loramendi molding line)	Machine: Molding Line, Masking Machine Material: Core; Method: work instruction, identification label; Person: Operator of the molding line	Production of a mold	Assembling the cores in the form of the castings necessary to recreate the internal shapes	Machine: Production of the casting mold; Material: Assurance quality of materials; Method: Presentation to the Employee of the quality requirements and the method of implementation, Identification of materials; Person: Operation implementation control	Occurrence of a non-compliant product (shape defects and sandiness of the castings)	8	Cracking of the cores in the casting mold during pouring the mold

(step 4)	**Risk analysis (step 5)**							**Optimization (step 6)**										
3. Cause of failure (FC)	Current control - prevention (PC) for FC	Occurrence (O) FC	Current control – Detection (DC) for FC or FM	Detection (D) FC/FM	PFMEA AP	Special characteristics	Codes	Preventive actions	Detection actions	Responsible person	Planned end date of the action	Status	Implemented actions with proof of implementation	The end date of the actions	Severity (S) FE	Occurrence (O) FC	Detection (D) FC/FM	PFMEA AP
Incorrect storage of the recently painted core	-	7	Visual inspection	8	H	-	-	1. Development of the core control system during the coating operation and the drying place	1. Introduction of a standard for surface verification	New project Manager	31.12.2021	CLOSE	According to plan	31.10.2021	8	1	8	L
Low temperature in the core room in winter	-	5	Visual inspection	8	H	-	-	1. Installing a gas dryer and introducing the standard of drying the cores in accordance with the drying time table before allowing the	-	Foundry Manager	31.10.2021	CLOSE	According to plan	31.10.2021	8	2	8	M 13
								cores to the line										

4 Conclusions

The growing importance of the implementation of the risk management process in enterprises has been influenced by the dynamic character of the environment, which has an immeasurable impact on the effectiveness of the organization. As forecast by most experts in the field of quality management, current trends of uncertainty in the organization's environment and the dynamism of changes taking place will only become deeper. Many qualitative and quantitative methods of risk assessment are used today, but most of them are based on the same steps, i.e. setting the context and identifying risk factors, analysing and evaluating risk. Nowadays, enterprises face the necessity of continuous improvement of the existing processes, as well as searching for new technical and organizational solutions in order to meet the applicable requirements [16].

The current edition of the ISO 9001 standard underlines the essence of understanding the context of the organization and implementing the management strategy in accordance with the risk-based approach. Adapting the concept of risk management in processes, in particular production processes, makes it possible to respond to the requirements of the interested parties and make informed decisions. It is necessary to make people aware and accept the fact that risk is an inherent link related to functioning on the market. Risk management should be a basic element built into the company's management strategy at every operational level. Popularizing risk management issues has contributed to the creation of many qualitative and quantitative methods supporting the implementation of the risk management process. One of the tools applied for risk assessment is the PFMEA analysis [15].

This article presents the use of the PFMEA analysis in production practice to detect the causes of defects, and on its basis, preventive and corrective actions were implemented in the production process on the moulding line. Ensuring high-quality products is one of the main goals of production plants. It is worth mentioning that the earlier a non-compliance is detected in the production process, the lower the costs of poor quality are associated with it. The PFMEA analysis is a tool that enables the prevention of any defects in the process and effective response in the event of their occurrence.

To sum up, thanks to the use of quality management instruments, consistent with the idea of continuous improvement, which include the PFMEA analysis, the occurrence of faults in both the production process and the product can be minimized by identifying the defects. The PFMEA analysis is an effective methodology for identifying problem areas and taking actions aimed at minimizing or completely eliminating the causes of inconsistencies in the process. It is worth noting that PFMEA also allows to prioritize activities and sets the direction for enterprises in the process of establishing an action plan, thanks to which it effectively ensures the required quality of foundry products [12, 17, 18].

References

1. Wróblewski, D.: Risk management. Review of selected methodologies, CNBOP-PIB, Józefów (2015)
2. PN-ISO 31000:2018, Risk management - rules and guidelines

3. LNCS Homepage. http://www.qualityaustria.com.pl/baza-wiedzy/art/rewizja-iso-9001-wyjasniona-w-prosty-sposob-kontekst-organizacji. Accessed 15 Oct 2021
4. LNCS Homepage. Zarządzanie ryzykiem w przedsiębiorstwie. https://repozytorium.uph.edu.pl/bitstream/handle/11331/822/Wroblewski_Zarzadzanie_ryzykiem_w_przedsiebiorstwie.pdf?sequence=1. Accessed 15 Oct 2021
5. LNCS Homepage. https://repozytorium.uph.edu.pl/bitstream/handle/11331/822/Wroblewski_Zarzadzanie_ryzykiem_w_przedsiebiorstwie.pdf?sequence=. Accessed 15 Oct 2021
6. LNCS Homepage. https://serwiszoz.pl/jakoscsanepid/jak-zarzadzac-ryzykiem-w-ujeciu-iso-90012015-2620.html. Accessed 15 Oct 2021
7. LNCS Homepage. https://fmea.com.pl/nowe-wymagania-aiag-i-vda-w-zakresie-fmea-definiowanie-wad-i-skutkow. Accessed 15 Oct 2021
8. LNCS Homepage. https://akademiajakosci.com/wp-content/uploads/2020/02/Nowe-FMEA-AIAGVDA-vs.-stare-FMEA.pdf. Accessed 15 Oct 2021
9. LNCS Homepage. https://inzynierjakosci.pl/wp-content/uploads/2019/06/magazyn_jakosc_11.pdf. Accessed 15 Oct 2021
10. LNCS Homepage. https://inzynierjakosci.pl/2019/05/fmea/. Accessed 15 Oct 2021
11. LNCS Homepage. https://automotivesuppliers.pl/pl/polska/rewolucja-w-fmea. Accessed 15 Oct 2021
12. Jamróz, A., Kostrzewa, N.: Optimization of the production process through the implementation of the SMED system in a printing company, In: Methodical and instrumental aspects of production engineering, AGH, Kraków (2014)
13. Hamrol, A., Mantura, W.: Quality Management - Theory and Practice. PWN, Warsaw (2011)
14. Perzyk, M.: Foundry. WNT, Warsaw (2012)
15. LNCS Homepage, Selected aspects of quality control in the foundry in terms of the requirements of ISO 9000. http://www.afe.polsl.pl/index.php/pl/3056/wybrane-aspekty-sterowania-jakoscia-w-odlewni-pod-katem-wymagan-norm-iso-9000.pdf. Accessed 15 Oct 2021
16. Czajka, K.: Assumptions of the process of organization of the network production system in the conditions of distributed resources, In: Methodical and Instrumental Aspects of Production Engineering, red. M. Dudek i in., AGH, Kraków (2014)
17. Brzozowski, L., Down, M., Younis, H.: FMEA Handbook. 4th edition, AIAG, Southfield (2004)
18. Dudek-Burlikowska, M.: Monitoring of the production processing in a metallurgical company using FMEA method. Arch. Metallur. Mater. **62**(4), 1-10 (2017). https://doi.org/10.1515/amm-2017-0309

Assessing the Emergency Assembly Areas Using Maximum Coverage Location Analysis: A Case of Gaziantep University

Eren Özceylan[1](✉) and Cihan Çetinkaya[2]

[1] Industrial Engineering Department, Gaziantep University, 27100 Gaziantep, Turkey
erenozceylan@gmail.com

[2] Department of Management Information Systems, Adana Alparslan Turkes Science and Technology University, 01200 Adana, Turkey
ccetinkaya@atu.edu.tr

Abstract. Emergency assembly areas (EAAs) are safe areas where the people can gather away from the dangerous area in order to prevent the panic that will take place until the temporary shelter centers are ready after disasters and emergencies. Determination of a suitable site for an EAA is crucial to decrease the negative impacts of disasters. There are a few criteria to be considered while finding a place for an EAA, e.g. assembly points should be located at a safe distance away from the danger (building, fire), they must be easily accessible, and finally they must be big enough to accommodate all potential victims. To solve this problem scientifically, the aforementioned conditions should be modeled as a maximum coverage location (MCL) problem. In this paper, the EAAs in Gaziantep University campus are discussed and evaluated. To do so, the 32 current points are considered as source nodes, and 65 buildings are considered as demand nodes. The covered population who are evacuated from buildings is maximized under different travel distance limits. An integer programming formulation is applied to evaluate the current EAAs and the suitability of existing signs is discussed. As a result, it has been determined that eight current EAAs are not suitable. According to another result, everyone can reach the remaining 24 EAAs within 196 m.

Keywords: Emergency assembly area · Evacuation · Integer programming · Location analysis · Maximum coverage location

1 Introduction

Emergency assembly areas (EAAs) are pre-determined areas, to be used in case of emergency situations. They provide a safe area for potential disaster victims to stand during the period for emergency aid. These safe areas also aim to prevent panic and ensure healthy information exchange in the period until temporary shelter centers are ready after disasters.

Not every empty space is suitable for EAAs. Various criteria are evaluated together while determining these assembly points. EAA determination criteria can be summarized

A. Hamrol et al. (Eds.): MANUFACTURING 2022, LNME, pp. 36–46, 2022.
https://doi.org/10.1007/978-3-031-00218-2_4

as; being away from secondary dangers, being accessible and finally having adequate place for keeping the population safe [1].

According to Turkish Disaster and Emergency Management Presidency strategic plan [2]: (*i*) increasing the effectiveness of coordination in disaster and emergency management and (*ii*) managing the processes during and after disasters in the most effective way are listed among the main objectives for disaster preparedness plan. In this context, determining the EAAs in strategic locations is crucial.

The problem area of this paper, namely Gaziantep city is the ninth most crowded city in Turkey [3]. In addition, as can be seen in Fig. 1, the city is determined among the most earthquake risky cities of the country.

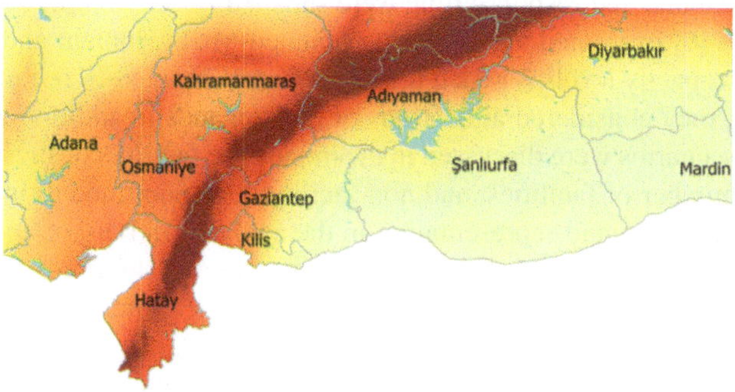

Fig. 1. The earthquake risk map of Gaziantep and its region [2].

Therewithal, the University of Gaziantep has almost 50 thousand people including students, academicians and staff in its campus, being among the most crowded institutions in the region. Thus it is important to make an emergency plan for the campus.

In this study, the EAAs in Gaziantep University campus are investigated. Firstly, the 32 current assembly points are considered as source, and 65 university buildings are considered as demand nodes. The number of potential victims is maximized regarding alternative distance limits. For the solution, an integer programming formulation is proposed to evaluate the current EAAs and offer new locations.

The outline of the paper is as follows; the next section examines the literature on MCL problems and defines how the proposed paper will fill the gap. The third part defines the proposed methodology and the fourth part includes the case study and analysis. In the fifth part, the summary of the paper is given and future directions are explained.

2 Literature Review

Since emergency management is a very important topic, there are a lot of papers with different approaches to similar problems. In this part, the emergency management studies that use the MCL analysis are summarized.

Araz et al. [4] proposed a multi-objective MCL model to solve the emergency service vehicles location problem. They aimed to determine the best locations for a group of vehicles while optimizing the service levels. For the solution of the problem, they proposed lexicographic multi-objective linear programming and fuzzy goal programming approaches. To prove the applicability of the proposed model, they provided numerical examples by using different solution approaches.

Balcık and Beamon [5] considered facility location problem for humanitarian relief chain. They developed a model to determine the number and locations of distribution centers and the total amount of supplies to be kept at each center. They formed the problem as a variant of the MCL model that integrates inventory decisions and facility locations. The computational experiments were conducted by using GAMS/CPLEX and they illustrated the efficiency of the proposed model on a real case problem. As a result, the effects of pre & post disaster relief funding on system's performance, especially on proportion of demand and response time were satisfied.

Yin and Mu [6] examined a modular capacitated MCL problem and formulated to allow different capacity levels for each facility at potential sites. To optimally site emergency vehicles, they considered allocations of the demands in addition to the covering standard. Two scenarios were discussed in the model: the facility constrained model that fixes the total number of facilities, and non-facility constraint model. In addition, they also applied spatial demand representation in the analysis and discussion. Finally, as a case study in the State of Georgia, Geographic Information System and optimization software packages were used together to site the ambulances optimally for the Emergency Medical Services.

Chanta et al. [7] focused on ambulance location problem to optimize the level of ambulatory services provided to patients in rural or urban areas. They modified the MCL models by proposing three bi-objective covering location models. The multi-objective problem was solved by the ε-constraint method. The obtained results were provided by using the data from Hanover County fire department, Virginia. A sensitivity analysis was applied to provide an insight on how the zone classification affects the solution. It was reported that the proposed method was effective and could help to reduce disparities in service levels in emergency management.

Zhang et al. [8] employed uncertainty theory to address the emergency service facility location problem under uncertainty. They made use of the uncertain location set covering model and also they examined MCL problem again in uncertain environment. In addition, they modeled the MCL problem by using different approaches, namely, the (α, β)-MCL model and the α-chance MCL model. Finally, their approaches were illustrated by a case study in Sichuan, China.

Paul et al. [9] tried to analyze the effectiveness of the current and optimal locations of emergency management assets that belongs to the Department of Defense in United States. They formulated a multi objective MCL problem and they developed a set of non-inferior solutions by making use of the "ε-constraint" method. By doing so, they offered several Pareto optimal decisions to the decision makers. They highlighted that, as a result of their case study; by the relocations determined by the model, additional 45 million people could be covered.

Li et al. [10] examined the ambulance location problem in emergency medical service by proposing a MCL model. They used improved double standard model for the solution of the problem. A real case study was conducted by using the emergency medical service data in Shanghai, China. It is reported that the proposed model improved the service levels and allocation of the ambulances.

Zhang et al. [11] examined the multimodal facility location problem in humanitarian logistics. The multimodality rooted in, landside, airside, air-ground transshipment and air-ground combined transport. They proposed a two-stage MCL model for the problem. A solution algorithm was proposed and computational experiment was conducted by using a Java platform and CPLEX solver. They validated the proposed model's efficiency in a case study of Beijing, China. It was reported that the proposed approach was suitable for large-scale disaster rescue.

Yang et al. [12] proposed a MCL problem often confronted in natural disasters. Their aim was to optimize the locations of the communication hub centers of the self-organizing mobile network that is established in fields not receiving signal. They formulated the problem with two mixed-integer mathematical linear programming models and two hierarchical objective functions; the shortest moved distance and the maximum coverage. A new linearization method was used to linearize and solve the proposed models optimally. In addition, they provided a MILP-based heuristic approach to solve larger problems. As the result of experimental analysis, they summarized that the proposed models could find good solutions and were applicable for the emergency rescue scenes.

Bahrami and Ahari [13] examined the MCL problem for emergency services in Iran. They formulated the proposed problem by M (t)/M/m/m queuing system which try to minimize the number of victims waiting in the queue for receiving services. The obtained results showed that the GAMS software was not able to solve large problems. So they proposed an NSGA-II algorithm to solve larger problems. They demonstrated the applicability of the model in real case study that took place in a metropolitan city in Iran.

Alizadeh and Nishi [14] presented a hybrid covering model that exploits the set covering and MCL problems. The problem that they took into account was locating the first aid centers in humanitarian logistic services in Japan. The dynamic set covering location problem determined the locations of the facilities and assigned the facilities and allocated demand points via dynamic modular capacitated MCL problem. They validated the model by a case study in Japan and the results were compared to other possible combined problem versions. As a summary, it was reported that their model provided better coverage percentages when compared to other covering models.

This paper will fill the gap in the literature by proposing MCL problem while modeling the distance, accessibility and expanse criteria. Also solving a real case problem of a very crowded institution namely Gaziantep University is another contribution of the paper.

3 Methodology

In this paper, the current locations of EAA signs are evaluated and also new locations for additional signs are discussed. To do so, one of the well-known location-allocation models namely maximum coverage location (MCL) model is used in this study.

The MCL problem considers the objective of locating a given number of facilities (EAA signs) to maximize the covered number of demand nodes (buildings include

students, staff, and academicians) and demand nodes are expected be covered entirely if nodes are in the range of critical distance of the facility (an EAA sign), otherwise it is assumed to be not covered [15]. The optimal result of a MCL problem depends on pre-decided critical distance, decision on a critical distance value without altering coverage may lead "fully covered" to "not covered" [16]. The formulation of MCL model applied is given as follows [17]:
Sets and indices

$i \in I$, set of demand notes (buildings),
$j \in J$, set of candidate facilities (sites of EAA signs),
S, the distance beyond which a demand point is considered "uncovered",
$N_i \{j \in J / d_{ij} \leq S\}$ the nodes j that are within a distance of S to node i.

Parameters

P number of facilities
d_{ij} shortest distance between locations i, j,
a_i population (students, staff, and academicians) to be served at demand node i.

Decisions Variables
x_j is 1 if a facility sited at the jth node ($j \in J$); 0 otherwise,
y_i is 1 if node i is covered by one or more facilities stationed within S; 0 otherwise.

$$\text{Objective Function maximize} \sum_{i \in I} a_i y_i \tag{1}$$

subject to

$$y_i \leq \sum_{j \in N_i} x_j, \tag{2}$$

$$\sum_{j \in J} x_j = P, \tag{3}$$

$$y_i, x_j \in \{0, 1\} \quad \forall i \in I, \forall j \in J. \tag{4}$$

The objective function (1) is maximizing the sum of covered population in the sets in which they are covered. Constraint (2) allows y_i to equal 1 only when one or more EAA signs are established at sites in the set N_i (that is, one or more signs are located within S distance units of demand point i). Constraint (3) states that exactly p amount of signs are to be located. The binary restriction is shown in Constraint (4).

4 Case Study

The campus of Gaziantep University is selected as the study area in this paper. The campus has a total of 2,248,301 square meters of open space. The 65 buildings (faculties,

Fig. 2. The locations of demand nodes (buildings).

departments, administration offices and etc.) are considered as demand points and shown (B1 to B65) in Fig. 2.

There are 32 EAA signs in the main campus at the time of this study. The locations of current signs (EAAs) are shown in Fig. 3. One of the current signs namely EAA24 (written in Turkish) is shown in Fig. 4.

Fig. 3. The locations of current EAA signs.

Fig. 4. The picture of EAA24 (written in Turkish).

To find dangerous zones, the heights of buildings are calculated. Magnitude of buffers is determined via buildings' heights and multiplying them by 1.5 [18]. These zones are debris coverage zones and EAAs must be stay away from debris. If a building is collapsed in an earthquake, the effect area of that building is shown in the Fig. 5. According to the Fig. 5, it is found that eight of the current EAA signs are located in the dangerous zones. Eight signs (EAA4, 11, 13, 16, 17, 24, 27, and 31) within the impact area are also shown in Fig. 5. It means that the mentioned signs are not located properly.

Fig. 5. The dangerous zones of buildings if they are collapsed.

To re-locate the missing signs and evaluate current signs, suitable areas where signs to be placed are determined. The blue areas that cover 8,002 square meters show potential available locations (see Fig. 6). As it is known, 0.75 square meter space is required per person in the assembly areas [18]. Therefore, a total of 10,669 people can gather in the areas shown in blue. University's main campus population is almost 50,000 including students, academicians, and staff. But not all these people are in the buildings at the same time. Therefore, the blue areas can cover one-fifth of the total population. Figure 6 also shows the proper 24 (32 – 8 = 24) current signs.

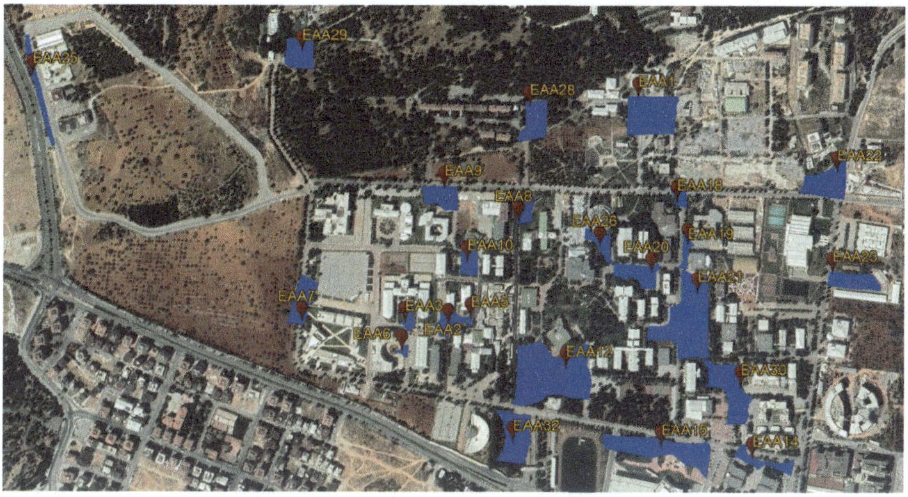

Fig. 6. Available emergency assembly areas.

At this stage of the study, maximum coverage model is applied. To apply the model, 24 signs are considered as a facility. Therefore, p is taken as 24. There are 65 buildings. 10,669 people are distributed to those 65 buildings (demand nodes) (see Table 1).

The mathematical model (Eqs. 1–4) is written in LINGO 14.0 using the data above. In the model, different coverage distances (S) are considered such as 25, 50, 75, 100, 125, 150, 175 and 196 m. The results of the model are given in Table 2 including the information of covered/uncovered buildings and population.

According to the Table 2, all people (10,669 people) can reach at least one EAA sign in 196 m. However, decreasing the coverage distance limit also decreases the number of covered people and buildings as expected. For instance, only 25.06% of people can reach an assembly area in 25 m. It must be noted that 24 of 32 current signs are considered in this analysis. Eight of them are located in unsuitable areas. Although 196 m is acceptable to reach an assemble area, all population can be covered in a shorter coverage distance if the eight idle signs are re-located.

Table 1. Population in buildings.

Building	Population	Building	Population	Building	Population	Building	Population
B1	15	B18	260	B35	12	B52	1
B2	6	B19	148	B36	295	B53	25
B3	3	B20	224	B37	15	B54	210
B4	4	B21	771	B38	5	B55	3
B5	323	B22	1	B39	5	B56	175
B6	1	B23	12	B40	5	B57	1
B7	101	B24	21	B41	342	B58	64
B8	121	B25	20	B42	218	B59	72
B9	68	B26	431	B43	432	B60	1
B10	531	B27	415	B44	14	B61	161
B11	149	B28	764	B45	25	B62	8
B12	4	B29	1532	B46	27	B63	5
B13	5	B30	595	B47	13	B64	4
B14	764	B31	1	B48	25	B65	1
B15	1	B32	17	B49	25		
B16	222	B33	190	B50	25		
B17	133	B34	1	B51	1	Total	10,669

Table 2. Results of MCL problem.

Distance limit	Covered population ratio	# of covered buildings	Uncovered buildings
25 m	25.06%	18	B2, B3, B5, B6, B7, B8, B4, B5, B17, B18, B19, B20, B21, B23, B24, B26, B27, B29, B31, B32, B33, B34, B36, B38, B39, B40, B41, B43, B44, B46, B47, B48, B49, B50, B51, B54, B55, B56, B57, B58, B59, B60, B61, B62, B63, B64, B65
50 m	66.12	37	B2, B5, B6, B14, B17, B24, B26, B27, B32, B34, B36, B38, B39, B40, B44, B46, B47, B48, B49, B50, B51, B54, B56, B57, B58, B62, B63, B65
75 m	76.38	50	B5, B14, B17, B24, B26, B32, B38, B39, B40, B46, B56, B57, B62, B63, B65
100 m	76.97	55	B5, B14, B17, B26, B32, B56, B57, B62, B63, B65
125 m	84.08	61	B5, B14, B32
150 m	84.20	63	B5, B14
175 m	91.36	64	B5
196 m	100.00	65	None

5 Conclusion

The EAAs are used in case of an emergency and the victims are kept there until partially or completely evacuation. In most of the governmental institutions, these assembly points are pre-determined and it is very important to inform the people about these points. The EAAs need to be clearly signposted to make it obvious; also these signposts can prevent others from parking there or dumping items to the area. Since our country – Turkey– had many disasters especially earthquakes for years, the Disaster and Emergency Management Presidency pays attention to the disaster education and preparedness. The year 2021 was determined as Disaster Education Year [19]. One of the most important objectives of this theme is determining the EAAs scientifically.

In this paper, the EAAs in Gaziantep University campus are taken into account. The distance, accessibility and the expanse of the area are chosen as the most important issues that affects the EAA location problem. Thus these conditions are modeled as a MCL problem. Current 32 assembly points in the campus are chosen as source nodes, and 65 buildings in the campus are considered as demand nodes. The maximization of covered population (students, academicians and staff) is aimed considering different travel distances. An integer programming formulation is applied to evaluate the current EAAs. As a result, eight of the 32 signs are found as unsuitable because they are located in the dangerous area. Therefore, the locations of eight signs should be changed. According to the model, the remaining 24 signs can cover all the population in 196 m. Decreasing the reach distance from 196 m to 25 m also decreases the covered population by 75%.

Consequently, a scientific approach is proposed for using in emergency management in the university campus. The results of the paper can be used by university administration and policy makers. For the future studies, geographic information system-based multi-criteria decision making techniques can be used together for determining the additional EAAs.

References

1. DEMP: Disaster and Emergency Management Presidency (2021). https://www.afad.gov.tr/toplanma-alanini-ogren-ki-canin-sag-olsun. Accessed 24 Nov 2021
2. DEMP: Institutional Reports: 2021 Performance Program (2021) https://www.afad.gov.tr/kurumsal-raporlar. Accessed 24 Nov 2021
3. TSI: Turkish Statistical Institute, Address Based Population Registration System Results (2021). https://data.tuik.gov.tr/Bulten/Index?p=Adrese-Dayali-Nufus-Kayit-Sistemi-Sonuclari-2020-37210. Accessed 24 Nov 2021
4. Araz, C., Selim, H., Ozkarahan, I.: A fuzzy multi-objective covering-based vehicle location model for emergency services. Comput. Oper. Res. **34**(3), 705–726 (2007)
5. Balcık, B., Beamon, B.M.: Facility location in humanitarian relief. Int. J. Logist. Res. Appl. **11**(2), 101–121 (2008)
6. Yin, P., Mu, L.: Modular capacitated maximal covering location problem for the optimal siting of emergency vehicles. Appl. Geogr. **34**, 247–254 (2012)
7. Chanta, S., Mayorga, M.E., McLay, L.A.: Improving emergency service in rural areas: a bi-objective covering location model for EMS systems. Ann. Oper. Res. **221**, 133–159 (2014). https://doi.org/10.1007/s10479-011-0972-6

8. Zhang, B., Peng, J., Li, S.: Covering location problem of emergency service facilities in an uncertain environment. Appl. Math. Model. **51**, 429–447 (2017)
9. Paul, N.R., Lunday, B.J., Nurre, S.G.: A multi objective maximal conditional covering location problem applied to the relocation of hierarchical emergency response facilities. Omega **66**, 147–158 (2017)
10. Li, R., Su, Q., Wang, Q., Wu, W.: A maximal covering location model of ambulances in emergency medical service. In: 15th International Conference on Service Systems and Service Management Proceedings, pp. 1–5. IEEE, Hangzhou (2018)
11. Zhang, M., Zhang, Y., Qiu, Z., Wu, H.: Two-stage covering location model for air ground medical rescue system. Sustainability **11**, 3242 (2019)
12. Yang, P., Xiao, Y., Zhang, Y., Zhou, S., Yang, J., Xu, Y.: The continuous maximal covering location problem in large-scale natural disaster rescue scenes. Comput. Ind. Eng. **146**(2), 106608 (2020)
13. Bahrami, I., Ahari, R.M.: A maximal covering facility location model for emergency services within an M(t)/M/m/m queuing system. J. Model. Manag. **16**(3), 963–986 (2020)
14. Alizadeh, R., Nishi, T.: Hybrid set covering and dynamic modular covering location problem: application to an emergency humanitarian logistics problem. Appl. Sci. **10**, 7110 (2020)
15. Karasakal, O., Karasakal, E.: A maximal covering location model in the presence of partial coverage. Comput. Oper. Res. **31**, 1515–1526 (2004)
16. Frade, I., Ribeiro, A.: Bike-sharing stations: a maximal covering location approach. Transp. Res. Part A Policy Pract. **82**, 216–227 (2015)
17. Church, R., ReVelle, C.: The maximal covering location problem. Pap. Reg. Sci. Assoc. **32**, 101–118 (1974). https://doi.org/10.1007/BF01942293
18. Spyridaki, P., Papadopoulou, I.D., Grigoriadis, V.N., Tziavos, I.N., Savvaidis, P.: Methodology for identification of emergency assembly areas and road network in cities using geographic information systems tools. In: International Multidisciplinary Scientific GeoConference Proceedings, Surveying Geology and Mining Ecology Management, Sofia, Bulgaria, pp. 135–142 (2009)
19. Bozkurt, Ö.: Disaster awareness culture and 2021 Turkey disaster education year (in Turkish). Turk. Adm. Assoc. **199**, 51–57 (2021)

The Triple Helix Concept in the Aspect of Counteracting Barriers in Science, Industry and the Public Sector

Jarosław Selech[1], Robert Sika[1](✉), Mariola Biedrzycka[2], Paweł Popielarski[1], Jakub Hajkowski[1], Andrii Zdobytskyi[3], and Mykhaylo Lobur[3]

[1] Poznan University of Technology, pl. Piotrowo 3, 61-138 Poznan, Poland
robert.sika@put.poznan.pl
[2] Partner of Promotion, Belgradzka 8, 02-793 Warsaw, Poland
[3] Lviv Polytechnic National University, 12 Bandery St., Lviv 79013, Ukraine

Abstract. The aim of this article is to present a set of barriers activities occurring in collaboration between scientists, business and infrastructure at the public sector level. The current capacity for the current strengthening of the knowledge networking and collaboration foundations in line with the concept of Triple-Helix (Academia - Business – Public sector) was highlighted. It is essential to transfer knowledge and technology from academia to industrial and social ground, highlighting and adapting solutions to some of the major problems related to industrial and societal challenges. The main goal is to create added value for the real economy. It is a particular challenge for companies with complex production technology (foundry, plastic processing, meld plastic working). In this respect, a special role lies in the good communication of the public sector, industry as well as modern universities focused on the willingness to cooperate. The examples of research results provide the financial commitment of particular triple-helix partners, which turned out to be successful in research and science.

Keywords: Triple helix · Innovation brokers · Counteracting barriers

1 Introduction

At the end of the 20th century, new opportunities appeared for a significantly accelerated development of production companies. This is due to the possibility of obtaining investment funds from the EU budget, which can be used in whole or in part to cover the costs of specific investments in fixed assets (machines and stationary devices), mobile measuring equipment or even commissioning research and development [1–5].

Many enterprises decided to introduce changes in the field of technology and work organization, which could make them more competitive in relation to other companies operating on the market. As you know, the competitiveness of enterprises is determined primarily by the possibility of satisfying customer needs in the best possible way. This means providing specific customers with the products or services that best meet their

A. Hamrol et al. (Eds.): MANUFACTURING 2022, LNME, pp. 47–59, 2022.
https://doi.org/10.1007/978-3-031-00218-2_5

requirements for timeliness, quality and price. Free market economy and hence – the dominant role of the client cause that these requirements are more and more diverse and individualized [6–8]. Recently, it has been especially noticeable in non-waste processes such as foundry, plastic processing, meld plastic working with assistive technologies such as laser technology or technology modeling [9].

The emergence of new business models based on innovation, collaboration networks and the expansion of endogenous resources is expected to make a strong contribution to the development of competitive European economies and regions. The triple helix [10, 11], following the logic of Smart Entrepreneurial Ecosystems [12] is not only based on a purely academic dimension, but also technological and entrepreneurial, in the most diverse sectors of activity.

Modern economics has reformulated conventional economic theory so that knowledge, technology, entrepreneurship and innovation are at the center of **the new model of economic growth,** rather than being seen as independent forces that are also largely independent of state policy [13]. The above is based on the fundamental assumption that the main objective of the state policy is to stimulate higher productivity and economic efficiency by stimulating innovation manifested in implemented projects. Consequently, it will also stimulate economic development. Nowadays, cooperation in complex innovative projects requires active cooperation and network connections with other entities and institutions.

Optimal cooperation between individual participants in projects can be ensured by **the triple helix (TH) model,** which is an innovation model that covers the mutual complex relationships occurring in the process of creating knowledge between three types of entities: research centers (universities, research centers, institutions supporting), industry (enterprises) and government (excluding local government institutions). The mutual relations between these three entities, undisturbed by barriers, determine the potential of cooperation. The lack of these connections significantly impedes the flow of knowledge [14]. Triple Helix describing the crossing of three worlds is presented in Fig. 1.

According to such a vision of an innovative economy, business is of key importance for all other activities, because both the networks of mutual connections in TH, the services provided and capital are focused mainly on entrepreneurs. Innovation intermediaries (brokers) are an important element in the network of mutual connections. They constitute a wide group of participants in innovation processes, knowledge exchange and technology transfer. Although they are not a party to these processes, they can perform many functions, such as foresight (discussion on the future among representatives of decision-makers – public authorities, scientific circles, industry, media, non-governmental organizations) or diagnosis of the demand for technologies, information and knowledge processing, gatekeeping (information dissemination) and brokering (technology flow control and brokerage), testing, building a framework for legal protection of commercialized knowledge, commercialization of knowledge and technology and many others [16, 17].

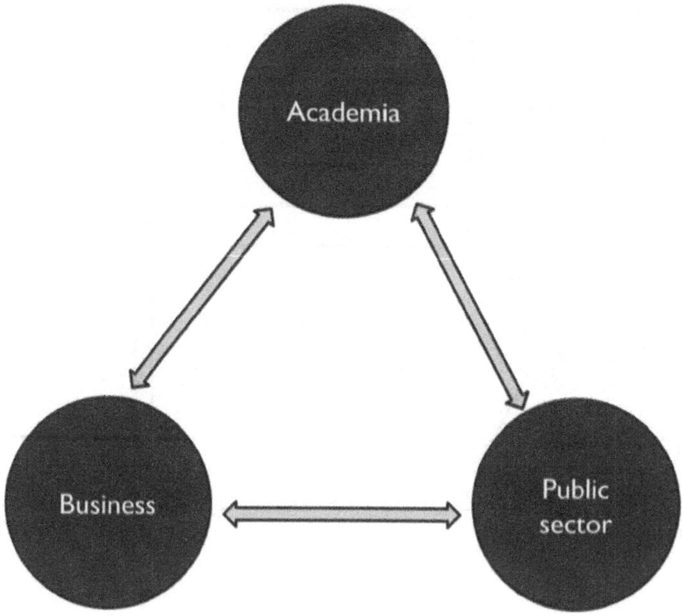

Fig. 1. The structure of mutual connections according to the Triple Helix assumptions [15].

For this, first of all, conscious, qualified staff – both on the side of government support institutions and on the side of BEI (a new model of compre-hensive service for the needs of innovative enterprises, so called INNO-BROKER), universities and entrepreneurs are need. Their goal is to provide specialist knowledge in the field of cooperation network management. Moreover, such a person is to be both an animator of relations between business and science, and a competence center in the area of e.g. intellectual property protection. The diagram of the relationship between the broker and other network participants within the TH is shown in Fig. 2.

Nowadays, science and technology play a very important role in technological progress and are one of the key driving forces behind the growth of modern economies. In order to effectively transfer knowledge from the science and research sector to the economy, appropriate tools and system solutions are necessary with participation. The most important of them, apart from legislative changes, include [18]:

- the existence in scientific and research centers of **units with appropriate human resources responsible for developing a network of contacts** between the world of science and business, and constant communication in order to exchange information taking into account the current demand for innovative solutions from industry or scientific achievements created in the units research,
- **intensification of cooperation** between university technology transfer centers (TTC) and business environment institutions, i.e. technology and industrial parks, business incubators, regional development agencies, scientific foundations or chambers of industry and commerce,

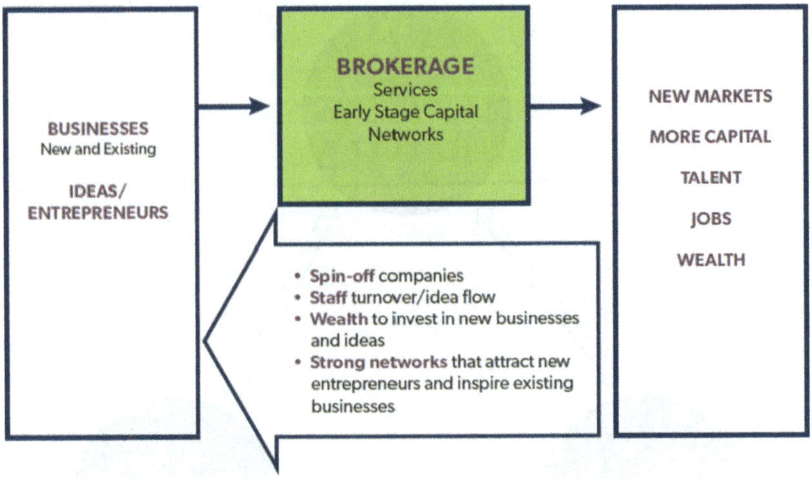

Fig. 2. The Innovation Broker in aspect of connects businesses and entrepreneurs to community resources [11].

- **creation of university platforms for the exchange of information** that would be used to disseminate the results of scientific works created at the university as well as provide the opportunity for the business sector to formulate inquiries and generate ideas directing the conduct of research works directly related to the needs of the industry,
- commitment of more funds to help researchers obtain funding from external sources for early stages of research, i.e. **proof-of-principle** and/or **proof-of-concept,** administered, for example, by government agencies, regional public funds or private funds characterized by high risk tolerance,
- **increasing the low awareness of the scientific** staff in the field of commercialization of knowledge and technology transfer, resulting from the unfavorable position of commercialization of scientific research results in the parametric evaluation system of both the university and the scientist,
- **support for activities in the field of technological brokerage** by assigning selected employees by universities and business environment institutions or by employing external experts specialized in specific areas of knowledge who, using effective mechanisms of verification and assessment of the market potential of conducted research works, are able to assess on an ongoing basis market suitability of the solutions being developed.
- Technology brokering may be performed by natural persons acting as a broker, technology broker, innovation broker or other entities. It is important that they can be both legal persons and organizational units that carry out similar activities and have appropriate competences. Technological brokerage carried out by scientific units primarily focuses on acquiring and analyzing information on the research and development work carried out at the university, while at the same time verifying the size of their

commercialization potential. In the event of a positive opinion on the commercialization potential of the created solution, the phase of presenting the obtained results among potential recipients of the developed technology takes place. The last step is to establish cooperation between the university and the business partner.

- The process of initiating the transfer of knowledge from universities to business carried out in this way, with the participation of specialists in the field of commercialization, is possible provided that **various barriers that arise between intersectoral partners are overcome.**

2 Tools and System Solutions Supporting Triple Helix

Nowadays, science and technology play a very important role in technological progress and are one of the key driving forces behind the growth of modern economies. In order to effectively transfer knowledge from the science and research sector to the economy, appropriate tools and system solutions are necessary with participation. The most important of them, apart from legislative changes, include [18]:

- the existence in scientific and research centers of **units with appropriate human resources responsible for developing a network of contacts** between the world of science and business, and constant communication in order to exchange information taking into account the current demand for innovative solutions from industry or scientific achievements created in the units research,
- **intensification of cooperation** between university technology transfer centers (TTC) and business environment institutions, i.e. technology and industrial parks, business incubators, regional development agencies, scientific foundations or chambers of industry and commerce,
- **creation of university platforms for the exchange of information** that would be used to disseminate the results of scientific works created at the university as well as provide the opportunity for the business sector to formulate inquiries and generate ideas directing the conduct of research works directly related to the needs of the industry,
- commitment of more funds to help researchers obtain funding from external sources for early stages of research, i.e**. proof-of-principle** and / or **proof-of-concept,** administered, for example, by government agencies, regional public funds or private funds characterized by high risk tolerance,
- **increasing the low awareness of the scientific** staff in the field of commercialization of knowledge and technology transfer, resulting from the unfavorable position of commercialization of scientific research results in the parametric evaluation system of both the university and the scientist,
- **support for activities in the field of technological brokerage** by assigning selected employees by universities and business environment institutions or by employing external experts specialized in specific areas of knowledge who, using effective mechanisms of verification and assessment of the market potential of conducted research works, are able to assess on an ongoing basis market suitability of the solutions being developed.

Technology brokering may be performed by natural persons acting as a broker, technology broker, innovation broker or other entities. It is important that they can be both legal persons and organizational units that carry out similar activities and have appropriate competences. Technological brokerage carried out by scientific units primarily focuses on acquiring and analyzing information on the research and development work carried out at the university, while at the same time verifying the size of their commercialization potential. In the event of a positive opinion on the commercialization potential of the created solution, the phase of presenting the obtained results among potential recipients of the developed technology takes place. The last step is to establish cooperation between the university and the business partner.

The process of initiating the transfer of knowledge from universities to business carried out in this way, with the participation of specialists in the field of commercialization, is possible provided that **various barriers that arise between intersectoral partners are overcome.**

3 Barriers in the Commercialization of Knowledge

3.1 Barriers on the Side of Scientists

In addition to the broadly understood system and organizational barriers related to the state and university policy, there are barriers related to the commercialization awareness and competences to carry it out, occurring in all entities involved in the commercialization process, i.e. the world of science, the economy sector and the business environment. On the side of the scientific community, the main barriers include [17, 19]:

- reluctance and low openness of scientists to commercialization activities and cooperation with the economy sector,
- lack or a small number of appropriate models for successful commercialization projects,
- no reaction of scientists to the emerging cooperation offers from industry,
- poor communication skills of scientists,
- reluctance to bear the risks associated with the commercialization of research,
- little knowledge of research workers about technology transfer mechanisms,
- imprecise and constantly changing provisions in the field of intellectual property law,
- too much bureaucracy in cooperation,
- lack of ongoing cooperation with the industry sector, which is mainly due to the limited flow of information about the specificity and details of activities and problems of entrepreneurs,
- concerns about understating the valuation of intellectual property,
- low supply of projects for commercialization.

3.2 Barriers on the Side of the University

The most important barriers identified on the side of universities include the reluctance of the scientific community to commercialization, which directly translates into reluctance to cooperate with Technology Transfer Centers responsible for knowledge transfer. Moreover, TTCs, which should be mainly responsible for the commercialization of technologies, often have too small and untrained staff to effectively carry out research commercialization processes, e.g. the lack of direct support from a patent attorney or lawyer on the part of the university during difficult talks with entrepreneurs [19, 20]. These difficulties result from the insufficient number of trained employees in relation to the demand. These difficulties are not only badly received by scientists, but also by entrepreneurs. In addition, the lack of developed mechanisms and basic knowledge about the available tools for the protection of intellectual property, often the lack of regulations and templates of agreements regarding intellectual property rights and broadly understood commercialization, licensing, conducting business activities by scientists in the form of spin-off companies or sharing profits from commercialization, result in the creation of barriers between science and business. Another barrier to commercialization is in many cases the low budget that universities have for basic commercialization activities, such as testing patent purity or technology valuation. Other barriers encountered in the process of commercialization of knowledge faced by universities include:

- no market demand for the developed technologies, which most often results from the mismatch between the offer of the obtained research results and the needs of the industry,
- lack of friendly formal and legal conditions,
- too few projects submitted for commercialization,
- complicated legal procedures related to technology transfer,
- low awareness among entrepreneurs about the possibility of establishing cooperation with the university.

3.3 Barriers of Entrepreneurs and Business Environment Institutions

The above-mentioned barriers were applied directly to scientists and universities, but it is not only scientists who are the main cause of weak links between universities and industry.

On the part of entrepreneurs, there are also identified limitations that adversely affect mutual relations, the most important of which are [17, 20]:

- lack of interest on the part of entrepreneurs in the implementation of research works by universities,
- low level of business confidence in the world of science,
- lack of wider interest of entrepreneurs in innovations,
- low legal awareness of enterprises,
- financial restrictions, especially for beginning SMEs,
- poor development of commercialization awareness,
- little experience in the field of cooperation with the world of science,

- uncertain return on investment in innovation or too long payback period,
- lack of a sufficiently effective legislative policy (unclear and non-transparent provisions, constant changes to them).

3.4 Barriers from Government Institutions

The most important institutional barriers to public administration include:

- legislative changes. Legislative changes include: amending the complex VAT Act, which causes difficulties not only for private entrepreneurs, but also for universities and other public institutions. It does not contain indications on how to introduce the invention to the market [19]. Consequently, the situation of the unit implementing the invention usually depends on many factors. In such a situation, it is necessary to apply for an individual interpretation of the provisions to the Tax Office, which makes it difficult to make many decisions, not only for accountants, because it is also associated with many months of waiting [20],
- regional policy not adjusted to the needs of entrepreneurs and scientists,
- public procurement law (PPL) not adjusted to the realities of universities. Polish public procurement law is not adapted to the realities of today's science [17]. The thresholds set out in the Act make it very difficult to conduct research, and the purchase of research equipment takes months and is not always successful. It is related to many formal and legal reasons that will not be discussed in this article,
- lack of a coherent model of introducing innovations to the market by universities. Many studies show that universities are not willing to establish cooperation with business [17]. There is still a sense of distance between the two worlds. The improvement of the situation requires the improvement of communication between the university or research and development unit and entrepreneurs,
- financing changes. In Poland, the path "from idea to market success" is very tedious and complicated. There is no need to convince anyone that, apart from a good idea, the key to the success of an innovative enterprise is finding an effective way to finance it in the initial phase.

It seems that in order to eliminate the above-mentioned barriers, profound institutional changes should be introduced, i.e.:

- legal order in the sphere of intellectual property protection,
- pro-innovative legal regulations in the science sector,
- improvement of the legal environment for the functioning of the economy in the field of innovation and technology transfer,
- creation of coherent regional innovation support systems,
- effective management of pro-innovative services.

4 Examples of Partnership Cooperation Implemented with Triple Helix

Below, this chapter describes in detail practical examples of cooperation carried out within the framework of ongoing projects, taking into account the triple helix formula [21, 22]:

- ISM project (see Table 1) [23],
- TPPA project (see Table 2) [24],
- KWB project (see Table 3) [25].

4.1 The ISM Project

The first of the ISM projects was entitled "The increase in the efficiency of the public transport as a result of implementing the concept of LCC and RAMS compatible to IRIS based on an integrated IT system" (see Table 1).

The final effects of the project were mainly used in the railway industry. As part of the project, an IT monitoring system was developed to collect, process and analyze data (see Table 1) on the damage to the fleet of rail vehicles (trams) according to the LCC (Life Cycle Costing) and RAMS (Reliability, Availability, Maintainability and Safety) concepts in accordance with the standard IRIS. The developed IT system was used by the manufacturer for quality management in the railway industry.

As part of the project, apart from the development of an IT system for the needs of Solaris Bus & Coach S.A., several master's theses were defended at the Poznan University of Technology, one habilitation thesis and many scientific publications.

The project consisted of a consortium, with an entrepreneur being the leader of the consortium, and the university acting as a consortium member.

Table 1. Data from the ISM project.

Design data				Name of the entity, role in the project and financial resources [PLN]		
Type of partnership	Acronym	Project duration [months]	Project budget [PLN]	Government	Education	Business
				National Research Center and Development	Poznan University of Technology	Solaris Bus & Coach S.A.
				Executive Founding Agency	Consortium member	Leader
Triple	ISM	26	3 080 000	1 780 000	–	1 300 000

4.2 TPPA Project

As part of the second, with the acronym TPPA (see Table 2), an innovative method of reducing the moisture content of loose materials in alternative fuels production technologies was developed.

On this basis, a prototype installation for reducing the humidity of an alternative fuel obtained from combustible waste in the form of a mixture of paper and plastic packaging was designed, built and tested, according to the patent registered in the Polish Patent Office under the number 397062. As part of the project, one invention was created, which received the claim Patent In addition, as part of the project, several engineers and master's degrees were created, which were defended at the Poznan University of Technology, as well as many scientific publications.

The second project, similarly to the first, consisted of a consortium, with an entrepreneur being the consortium leader, and the university acting as a consortium.

Table 2. Data of the TPPA project.

Design data				Name of the entity, role in the project and financial resources [PLN]		
Type of partnership	Acronym	Project duration [months]	Project budget [PLN]	Government	Education	Business
				National Research Center and Development	Poznan University of Technology	Ltd. Ekopoz
				Executive Founding Agency	Consortium member	Leader
Triple	TPPA	24	4 113 000	3 290 400	–	822 600

4.3 The KWB Project

As part of the KWB project (see Table 3), a prototype technological line was developed to obtain humic acids or potassium humate from lignite on an industrial scale and to obtain an innovative product in the form of humic acid with established parameterization values for the product ensuring its high quality.

Process innovation: an innovative technology for the production of humic acids from lignite on an industrial scale. As part of the project, one invention was created and several engineering and master's theses, one doctoral thesis and many scientific publications were created. Moreover, there were trainings and demonstrations of new technology for students.

In this project, the main contractor of the project was selected, which was business, while the universities acted as subcontractors. Summarizing the above-described examples of partnership cooperation implemented as part of the triple helix, the successor of its positive effects, expressed by project indicators, can be mentioned:

- joint patents, publications and conferences published in popular scientific journals with high impact factors,
- completed engineering, master's, doctoral and habilitation theses with very good final grades,
- joint technology development as part of basic and industrial research and development works,
- employing research workers in enterprises, which took place in the third of the discussed projects,
- special training, demonstrations and demonstration seminars for a specific technology.

However, the most important of the effects is jointly developed technologies that will contribute to the increase in the competitiveness of many enterprises and the overall dimension to technological progress and increase in the innovativeness of the economy.

Table 3. Data of the KWB project.

Design data				Name of the entity, role in the project and financial resources [PLN]		
Type of partnership	Acronym	Project duration [months]	Project budget [PLN]	Government	Education	Business
				National Research Center and Development	Poznan University of Technology, University of Technology and Life Sciences in Bydgoszcz	Brown Coal Mine Ltd. "Sieniawa"
				Executive Founding Agency	Subcontractor	General contractor
Triple	KWB	26	23 818 884	14 915 078	–	8 903 806

5 Conclusions

As indicated above, universities often do not have separate or well-specialized units for cooperation with industry with sufficient human resources. Therefore, in order to improve the efficiency of the process of commercialization of knowledge or greater integration of science with industry, the barriers that have arisen should be eliminated. The aforementioned technology brokerage, involved not only in the industry, but also in leading research centers, can help.

The key factor deciding on overcoming some barriers is the broker or broker using both formal and informal contacts through a network of connections and personal acquaintances, as well as reaching the sources of information about research projects and the results obtained. This may be a strong support for CTT in the process of commercialization of knowledge, as the information obtained often concerns projects which managers have not submitted for commercialization for many reasons. In addition, the form of personal contact with the scientist allows you to quickly assess the commercial potential of the works carried out by the scientist, as well as translates into their later more understandable presentation to the interested entrepreneur. As shown in the given examples, the government is able to provide a large share of the required funds with an appropriate consultations between the university and industry, taking into account the industry's own financial contribution.

Acknowledgments. The presented results were funded by grants for education allocated by the Ministry of Science and Higher Education in Poland.

References

1. Mousavi, S., Bossink, B., Vliet, M.: Microfoundations of companies' dynamic capabilities for environmentally sustainable innovation: case study insights from high-tech innovation in science-based companies. Bus. Strategy Environ. **28**, 366–387 (2019)
2. Sika, R., et al.: Decision support system in the field of defects assessment in the metal matrix composites castings. Materials **13**(14), 3552-1-3552–27 (2020)
3. Ludmyla, Z., Petro, M.M., Chepurda, G., Lyzunowa, O., Shmygol, N.: Strategy of innovative development as an element to activate innovative activities of companies. Acad. Strateg. Manage. J. **18**(4), 1–6 (2019)
4. Zdobytskyi, A., Mykhailo, L., Panchak, R., Sika, R., Kalinowski, K.: Increasing the strength of materials by topological optimization methods. In: IEEE 16th International Conference on the Experience of Designing and Application of CAD Systems (CADSM), vol. 21, ISBN 978-1-6654-3894-0 (2021)
5. Popov, G., Legutko, S., Kliemntov, K., Kostov, B.: Applying criteria equations in studying the energy efficiency of pump systems. Energies **14**(17), 5256-1-5256–13 (2021)
6. Srinivasan, P.L., Kadu, K.: Organizational change with respect to technology and workforce for creating an optimized organization. Int. J. Adv. Res. Manage. **3**(9), 181–186 (2018)
7. Bartkowski, D., Bartkowska, A., Piasecki, A., Jurci, P.: Influence of laser cladding parameters on microstructure, microhardness, chemical composition, wear and corrosion resistance of Fe–B composite coatings reinforced with B4C and Si particles. Coating **10**(9), 809-1-809–18 (2020)
8. Matysiak, M., Duspara, M., Cumin, J., Smardzik, I.: Conventional shaping the edges of thick steel plates after welding process. Techn. Gaz. **28**(2), 663–667 (2021)
9. Sika, R., Hajkowski, J.: Synergy of modeling processes in the area of soft and hard modelling. In: 8th International Proceedings on Manufacturing Science and Education – MSE, MATEC Web of Conferences, Sibiu, 050071–050078 (2017)

10. Galvao, A., Mascarenhas, C., Marques, C., Ferreira, J., Ratten, V.: Triple helix and its evolution: a systematic literature review. JST **10**(3), 812–833 (2019)
11. Scalia, M., Barile, S., Saviano, M., Farioli, F.: Governance for sustainability: a triple-helix model. Sustain. Sci. **13**(5), 1235–1244 (2018). https://doi.org/10.1007/s11625-018-0567-0
12. Gorelova, I., Dmitrieva, D., Dedova, M., Savastano, M.: Antecedents and consequences of digital entrepreneurial ecosystems in the interaction process with smart city development. Adm. Sci. **11**(3), 94, 1–14 (2021)
13. Oettinger, J., Henton,, D.: The role of innovation brokers in a knowledge economy: the fourth strand to triple helix. In: Collaborative Economics, paper submitted to the Triple Helix XI International Conference, London (2013)
14. Bojar, M., Machnik-Słomka, J.: Model of a triple and quadruple helix in building network cooperation for the development of innovative regional projects, science notebooks in Silesian University of Technology. Organ. Manage.. **76**(1923), 99–111 (2014)
15. Beyond "Triple Helix" – towards "Quad Helix." The Bearing Wave, Bearing Consulting Ltd. March (2022)
16. Abbas, A., Avdic, A., Xiaobao, P., Hassan, M.M., Ming, W.: University-government collaboration for the generation and commercialization of new knowledge for use in industry. J. Innov. Knowl. **4**(1), 23–31 (2019)
17. Leydesdorff,, L., Etzkowitz, H.: The transformation of university-industry-government relations. Electron. J. Sociol. http://arizona.openrepository.com/arizona/handle/10150/106531. Accessed 06 Feb 2018
18. Cai, Y., Etzkowitz, H.: Theorizing the Triple Helix model: past, present, and future. Triple Helix **7**, 189–226 (2020)
19. Champenois, C., Etzkowits, H.: From boundary line to boundary space: the creation of hybrid organizations as a Triple Helix micro-foundation. Technovation **76–77**, 28–39 (2018)
20. Leydesdorff, L.: Synergy in knowledge-based innovation systems at national and regional levels: the Triple-Helix model and the fourth industrial revolution. J. Open Innov. Technol. Mark. Complex **4**(2), 16, 1–13 (2018)
21. Zhang, Y., Chen, K.: Scientific effects of Triple Helix interactions among research institutes, industries and universities. Technovation **86–87**, 33–47 (2019)
22. Anttonen, M., Lammi, M., Mykkanen, J., Repo, P.: Circular economy in the Triple Helix of innovation systems. Sustainability **10**(8), 2646 (2018)
23. Research project "The increase in the efficiency of the public transport as a result of implementing the concept of LCC and RAMS compatible to IRIS based on an integrated IT system", PBS3/B6/30/2015. https://www.solarisbus.com/pl/firma/dotacje. Accessed 27 Oct 2021
24. Research project "Development of Innovative methods of lowering of moisture content granular materials in alternative fuels production technology", GEKON-213086, 2015–2016, https://ekopoz.eu/projekty.html. Accessed 27 Oct 2021
25. Research project "Investigating the property of lignite and assessment of its suitability in production of humic acid", POIR.01.01.01-00-0799/16-04, 2017–2020. http://www.sieniawa.com/dofinansowanie. Accessed 28 Oct 2021

Total Quality Management: Practices to Leverage Its Principles in Distance Higher Education

Márcia Helena Borges Notarjacomo, Bruna Strapazzon Do Couto(✉), Fernanda Bica de Almeida, Miriam Borchart, and Giancarlo Medeiros Pereira

UNISINOS University, Avenue UNISINOS, RS 950-93022-000 São Leopoldo, Brazil
brunascouto9@gmail.com

Abstract. The growth of distance learning in higher education has been gaining market share. The COVID-19 pandemic has reinforced this trend. Enrollments are increasing sharply while fierce competition is a threat to distance learning institutions. To conquer and retain distance learning students, it is necessary to offer quality services and products, seeking innovation and continuous improvement to meet their needs. However, studies that analyze the application of the principles and practices of TQM (Total Quality Management) in distance higher education are still scarce. From this perspective, this study seeks to identify practices that increase the presence of TQM principles in distance higher education institutions. The qualitative research was carried out with managers from 64 teaching centers of one of the largest distance learning institutions in Brazil. The pillars of TQM considered in this study are leadership, staff, students, technological resources and continuous improvement. This study contributed to the literature by identifying practices to expand the presence of TQM principles in distance higher education adopted by the studied institution and its teaching centers. Management contributions are also presented.

Keywords: Distance learning · Higher education institutions · Total quality management · Distance higher education · Service quality

1 Introduction

This research considers three main aspects. The first is related to the relevance of distance learning in higher education. Distance learning has received increasing attention from the academic field as well as society in general [1]. The popularization of the internet has contributed to the spreading of distance learning to remote regions around the world [2]. Additionally, the COVID-19 crisis has accelerated the offering of distance learning by colleges and universities exacerbating the challenges and advantages to both: providers of service - college and universities - and students [1]. These authors reinforce the importance of understanding the impact of distance learning on the effects of education and the social consequences of this type of education.

A. Hamrol et al. (Eds.): MANUFACTURING 2022, LNME, pp. 60–71, 2022.
https://doi.org/10.1007/978-3-031-00218-2_6

The second aspect is related to the quality of the distance learning education service. Dynamic market forces are compelling higher education institutions to adapt to remain relevant and competitive [3]. In this context, quality is an essential element to contribute to improving internal processes while leveraging student loyalty. To the author, in the context of higher education, quality refers to fitness for purpose, value for money, perfection, transformation and distinctiveness, and is categorized in terms of educational quality and administrative quality. Total Quality Control (TQM) is considered as one of the critical determinants to support organizations in improving their financial and operational performance [4]. TQM is a systematic approach for the continuous improvement of all organizational processes through total participation of all employees, resulting in high-quality products and services to attain customer satisfaction. The philosophy of TQM is "Do the right things, right the first time, every time" [5].

The third aspect considers that both themes separately – distance learning in higher education and TQM – have been widely discussed in academic literature. However, when distance learning is combined with TQM, there are still few academic studies [6]. It is also noteworthy that implementation of TQM in services is made difficult by the characteristics of services [7]: service heterogeneity, intangibility, and perishability, as well as the simultaneity of production and consumption, cause more complex applicability of quantitative tools and techniques of TQM [8].

In the next section, it is presented the research problem. After, the literature review is presented, concerning principal concepts related to TQM, TQM in higher education institutions, and TQM pillars in distance learning. Next, the method is presented. Then, the results and discussion. In the end, conclusions are exposed.

2 Research Problem

Effective service quality measures for distance learning have been required to evaluate the satisfaction of students. It is essential to distance learning offers to avoid evasion because students not satisfied with the course easily find other options in the market [9]. The satisfaction of students is a consequence of distance learning quality that can be leveraged by the application of TQM principles [6]. TQM was used as an analysis lens for this research because it is a consolidated structure, widely applied in services and focused on continuous improvement [7]. Understanding what TQM principles contribute to improve the distance learning quality and how they are implemented are understudied in academic literature [6]. Such understanding of what principles and how to apply them are relevant concerning TQM in distance learning offers because the COVID-19 pandemic environment has accelerated the offering of distance learning courses around the world [1]. Educational institutions focus on quality of their products and processes, supported by well-structured quality system, possibly will achieve better performance and satisfaction of students.

The presented context of distance learning and the need to expand the application of TQM in higher education institutions led us to the question: How can TQM principles be incorporated in higher education institutions that only offer distance learning courses? The consolidation of TQM principles and practices could help the institutions to remove inefficiencies, keep the focus on the market demands and threats, achieve better performance in all areas, and satisfy stakeholders' (students, partners, teachers, etc.) needs. To understand how TQM is implemented, a case study was performed in one of the biggest distance learning higher education institutions in Brazil. The managers of 64 educational centers were interviewed, and practices and documents were analyzed.

This study contributes to the literature by proposing a framework to TQM principles in distance learning in higher education and identifying practices adopted by the studied institution and its educational centers. Managerial contributions are presented.

3 Literature Review

3.1 Total Quality Management (TQM)

Academic literature has a multitude of studies on Total Quality Management (TQM) and its applications in the most diverse sectors. Many authors disagree on the correct definition of TQM, as there are different variables and principles to be considered. Older studies bring a broad view of the various aspects that involve TQM, such as its concepts and definitions, the roles assigned to senior management leadership, employees, the tools and methods used [10–13], as well as the importance of a quality-oriented organizational culture [4]. In a more modern view, TQM practices are understood as strategic tools, widely used by organizations to generate competitive advantages through performance optimization [14].

Briefly, the TQM is consolidated through concepts and techniques that aim to boost competitiveness through the continuous improvement of processes, involving the permanent and long-term management of all organizational resources and stakeholders involved [15]. As it encompasses the entire management of the organization, the TQM approach proposes a new management model that seeks to make organizations more efficient and flexible [16].

The so-called TQM pillars are essential elements for the implementation of a structured and strategically designed system, as the implementation of good management practices requires changes in the organization's internal arrangements [17]. There are theoretical divergences as to the principles that guide the composition of TQM. Some researchers point to 10 topics based on four different studies. They are: senior management support, customer relations, supplier relationships, workforce management, employee behavior, product development, process management, quality reporting and data, the role of the quality team and benchmarking [18]. Other authors list eight principles considered key elements for the development of TQM [19, 20]. In general, there are: (I) customers; (II) leadership; (III) continuous improvement; (IV) systemic approach to management; (V) mutual benefits in relations with suppliers; (VI) management by processes; (VII) decision-making based on facts and data; and (VIII) employee involvement.

In general, there are: (I) customers; (II) leadership; (III) continuous improvement; (IV) systemic approach to management; (V) mutual benefits in relations with suppliers; (VI) management by processes; (VII) decision-making based on facts and data; and (VIII) employee involvement.

3.2 TQM in Higher Education Institutions

The growth of competitiveness in service organizations has been favored, especially in recent years, by the application of operational excellence tools, with a focus on quality improvement, cost reduction and process acceleration [21]. Faced with a dynamic and increasingly competitive educational scenario, valuing the improvement of service quality is an essential prerequisite for higher education institutions, requiring the reformulation of strategies for an emerging market of mass services [22].

Quality management and improvement initiatives are continuously established in all service delivery industries, as these initiatives are aimed at meeting the needs and desires of users of the services offered [6]. As in other service-providing industries, the education sector also aims to meet the needs and desires of its users [23]. About quality management in higher education institutions, the main constructs are related to the quality of student learning, their involvement with the institution, quality and satisfaction in the provision of services, TQM, quality assurance, benchmarking and accountability [24].

In TQM's philosophy, the concept of quality will not be defined by the service that the institution expects to offer, but rather by the expectations of its clients. Organizations that follow the TQM path see quality as being defined by their customers. They are the ones who make the final decision on the perceived quality and, without them, the institution would not exist. Thus, the institution that adopts the TQM philosophy must use all available means to explore the needs of its clients [25]. The implementation of TQM, in the area of education, can shape educational institutions to meet the needs of different stakeholders, such as students, parents, the market and society in general [13].

The study by Tan, Muskat and Zehrer [26] identified five major research streams present in the literature: the exploration of the student experience; exploration of the learning experience; gender differences in the evaluation of experience in higher education; improving the quality of the student experience; and student satisfaction with the higher education experience. The study by Mehta, Prakash and Nitin [6] suggests thirteen guiding principles for the implementation of TQM in the education sector: (I) Institutional resource management; (II) Long-term strategy and planning; (III) Excellence HRM; (IV) Continuous assessment and improvement; (V) Top management commitment and visionary leadership; (VI) Student focus; (VII) Employee focus; (VIII) Alumni focus; (IX) Information management system; (X) Quality mission and vision statement; (XI) Service culture; (XII) Innovative academic philosophy and method; and (XIII) Industry institution partnership.

The principles listed cover a variety of technical and behavioral topics to be worked on, which include strategic planning, continuous improvement and a focus on internal and external customers, represented by the institution's employees, students and alumni.

3.3 TQM Pillars in Distance Learning

From a literature review, 11 articles were selected in order to identify the main TQM pillars and how they best relate to distance learning [2, 6, 24, 27–34]. Thus, relevant elements for the approach of each of these pillars were identified. In this way, a preliminary theoretical basis is elaborated, which serves as the starting point for the empirical research of this study.

From these related studies, it is identified that leadership, staff, students, technological resources and continuous improvement should govern the management of such educational institutions. Four studies indicate that leadership, combined with good planning, builds relationships, and improves the quality of service provided by the institution [24, 27, 29, 31]. All authors list the training and monitoring of professionals as a key aspect for standard teaching quality, encouraging their high performance and, consequently, minimized costs and times for preparation and execution of services. Seven studies list student-centered teaching as essential for the quality of distance learning, which includes rapid feedback and assistance to them, as well as constant monitoring and measurement of their dissatisfaction, so that there is an understanding and improvement of such aspects [2, 6, 24, 27, 28, 30, 34]. Four studies show that technological resources should be used in favor of more fluid, agile and complete teaching, in order to provide access to classes and material at any time, the exchange of information between students and tutors, in addition to enabling familiarization of students with the professional environment from virtual reality [24, 28, 32, 33]. Furthermore, all studies indicate continuous improvement as a pillar of distance learning, mainly as a result of employee training, the pursuit of innovation and prominence, and adaptation according to the needs of students. The results of this review are detailed next.

Regarding the pillar leadership, management and leadership operate at different levels because while the manager looks for the best ways to solve problems, the leader determines what exactly needs to be done [29]. Few publications in the current literature bring complete and structured studies on leadership as a fundamental part of teaching and learning processes [24].

The leadership principle is directly related to decision-making processes. To manage is to make the right decisions, even if they do not satisfy most stakeholders involved in learning [31]. In the education service provision environment, leaders influence and define instructional activities and processes, academic and social support, and the hiring and development of employees, establishing what strategic priorities should be for the use of available resources [27]. In this way, leaders build relationships and improve the quality of the service provided by the institution.

With regard to educational management, a leader will act as a mentor, serving clients - students -, at the administrative level, seeking to meet needs, solve problems and doubts that may arise. In some cases, there may be a leadership team, with the task of finding better ways of teaching [24]. In carrying out their daily activities, leaders promote instrumental goals, while recruiting faculty for the change effort, they cultivate a growing cadre of leaders in this process [27].

Regarding the pillar students, the quality of a service provided must always be improved, evaluated and measured reliably. Regarding educational institutions, it is essential to apply solutions for measuring quality developed from the students' point of view [35]. Student focus depends on a number of other factors, such as improving the overall performance of faculty, staff, benchmarking, assessments, safety, skills, alumni feedback, syllabus, partnerships, and more. These factors represent the critical outcome, which should receive special care from senior management [6].

Baig, Abrar, Ali and Ahmad [5] emphasize the importance of tolerance to freedom of expression and partial participation in decision-making processes within the institution by the student. This is so that they feel included in the teaching environment. The standardization of teaching must be combined with personalized service to each student, so that they acquire a feeling of belonging and importance, increasing their satisfaction and achievement.

Concerning pillar staff, it is reinforced that an educational institution is an organization that requires a lot of human resources. The training of the faculty is one of the essential factors for the quality of the service delivered. This aspect requires special attention from leaders with regard to the ability to recruit and retain trained employees, the effectiveness of performance feedback and professional development, and social resources within work teams [27].

Baig, Abrar, Ali and Ahmad [5] emphasize the importance of training and monitoring teaching professionals for the standardization of the service, while they should have the freedom to plan the time, place and duration of classes. The authors also emphasize that the participation of professionals is encouraged in the innovation of the teaching method, in order to explore modern and digital tools so that quality and teaching time can be gained. For Gay and Betts [30], the humanization of these professionals becomes essential for the quality of the service provided; thus, they reveal the importance of personalized communication, inclusion, equalization, follow-up, and quick feedback to students.

It is important to emphasize that, in addition to teaching professionals, the administrative and practical sectors of institutions must also retain some attention, as their correct execution guarantees a better experience for customers [2]. According to Gay and Betts [30], the correct handling of service in the first instance is capable of building customer loyalty, even before they use the service. Furthermore, the technical support team must be agile so that students do not miss out.

Regarding the pillar technological resources, distance learning is designed to promote independent learning, which eliminates the need for students to be physically present in the same environment as peers and teachers [24]. Such a system uses electronic technologies to enable the interaction of students and teachers separated geographically, temporally, or both [33]. Technological resources must be employed so that the student feels comfortable so that their learning is more effective. This also includes the infrastructure and study environment, which must be inclusive, secure, standardized, easily accessible and student-centered [27]. Gay and Betts [30] add that the use of technological resources such as virtual reality allows students to have theoretical and practical learning.

The availability of technological resources makes the educational process more accessible, as the asynchrony of classes makes the schedules and pace of learning more flexible, resulting in student satisfaction in relation to educational processes [36]. However, Tan, Muskat and Zehrer [26] emphasize that the digital platforms used must be easy to use, so that users can enjoy agile and punctual service and research. Still, it is suggested the provision of training, especially for students not yet familiar with the technology. Gay and Betts [30] also point out the importance of a fluid system and the provision of continuous support so that the experience does not become negative.

Finally, regarding the pillar continuous improvement, with the expansion of the offer of higher education courses and the increase in competition among higher education players, it is increasingly important that higher education institutions pay attention to the quality and perceived value of their services, both for managerial and business reasons, as well as the requirement established by Organs regulatory bodies [37]. Higher education institutions have faced new challenges every day, due to the dynamism of a scenario of continuous changes and full of demands, characteristics of the current educational market. Thus, when talking about quality and continuous improvement with regard to education, it is necessary to consider how it is understood and evaluated [38].

4 Method

This is a case study operationalized through qualitative research [39], in which the concepts and definitions of Total Quality Management in higher education institutions and the pillars of TQM regarding distance learning were addressed.

In order to establish a starting point for the TQM framework in distance learning, we seek to identify how the topic has been addressed in the literature. From the search in the databases, exclusion criteria were established in order to select only articles that addressed applications or literature reviews in relation to the application of the TQM in distance higher education environments. The exclusion criteria adopted were: (i) duplicate articles between the two bases, (ii) complete reading of the abstracts and (iii) complete reading of the articles to verify the elements related to each of the main pillars for the application of the TQM in teaching from a distance.

To better understand the current scenario of distance learning in higher education institutions, interviews were conducted with managers of educational centers from one of the largest distance learning higher education institutions in Brazil. Direct observations and documentary analyze were also carried out in part of the educational centers. Documents such as internal procedures, performance indicators, part of the strategic planning were analyzed. These interviews sought to identify common relevance factors in different educational centers to draw a profile of the practices and presence of the TQM pillars and understand the concerns and strategies adopted in terms of quality management in this sector.

For this, managers of educational centers from different Brazilian locations and of different sizes were chosen. In total, 30 interviews were carried out with managers from 64 educational centers, distributed in the five regions of Brazil. Some managers coordinate more than one center. Some managers coordinate more than one center. The higher education institutions studied have 300,000 active students and have already graduated 100,000 students in more than 60 different undergraduate courses. It also offers postgraduate courses, has 900 educational centers in Brazil and 5 abroad teaching in Portuguese.

In addition to the distribution by region, an attempt was also made to apply the research to educational centers of different sizes, in order to better capture the differences between them. Thus, managers of 49 small-sized centers, 10 medium-sized centers and 5 large centers were heard.

To carry out the interviews, a questionnaire was prepared with 10 open questions. The questions are related to the pillars identified in the literature: leadership, staff, students, technological resources and continuous improvement. The objective was to understand present aspects of each pillar and how they have been applied from the perspective of each interviewee. 4 interviews were face-to-face and 26 took place remotely. All were recorded and transcribed.

For the organization of data and analysis of the results, the answers of each interviewee were organized in a table in which the TQM pillars considered were included. In this way, the convergences and diverses perspectives of the respondents could be analyzed. Furthermore, for the proposition of the final framework, the pillars identified in the literature were considered, complemented with the practices pointed out in the interviews and with the documents made available by the institution studied.

5 Results and Discussion

Regarding the TQM elements in distance learning, the literature has several references to the importance of continuous improvement strategies, from a strategic and service management point of view. The points highlighted as pillars in the theoretical framework are due to literary findings, as in the study published by Mehta, Prakash and Nitin [6], in which principles are identified to guide the implementation of the TQM in the educational area.

The authors frequently highlight the importance of strategic leadership, responsible for institutional decision-making. Bryk, Sebring, Allensworth, Easton and Luppescu [27] state that, in the academic environment, leaders influence and define instructional activities and processes, academic and social support, and the hiring and development of employees, establishing what strategic priorities should be for the use of available resources. In this way, leaders are responsible for building relationships and, consequently, for the quality of the service provided by the institution.

After completing the interviews, it was evidenced that the higher education institutions addressed have a well-defined standard, applied and controlled in all their partner educational centers. As much as the educational centers have some independence in their decision-making, they must still provide a level of service required by the head office. This working method has been successful in partner educational centers, regardless of size, location or courses offered. This standardized work is possible due to the support offered by the head office, whether in the form of staff training, instruction for the accommodation of the structure and assistance in strategies in pursuit of the achievement of goals.

It is noticed that the larger educational centers have a management model that is a little different in relation to employees. In large educational centers, the coordinator leads the managers of each physical structure, while in small educational centers, this leadership comes entirely from the manager. The achievement of funding targets, in turn, proved to be the main objective of all interviewed managers, precisely because students are the source of the business.

The satisfaction of the student that has already been recruited is the next objective, as maintaining the acquired base is as important as increasing it with each module - the period for attracting new students. This satisfaction is periodically measured through surveys by the CPA (Proper Assessment Commission), a requirement of the Ministry of Education in Brazil, to measure the quality of service provided by higher education institutions. The results obtained through these assessments, in addition to serving as a basis for the accreditation and de-accreditation of higher education courses, also guide managers in promoting continuous improvement in their courses, in the infrastructure offered and in procedures adopted.

Based on the insights obtained, a framework is proposed that presents the pillars of TQM with regard to distance learning. From the conclusions obtained through interviews carried out with area managers, the main topics to be worked on in each of the five principles initially defined were pointed out, as observed in Fig. 1.

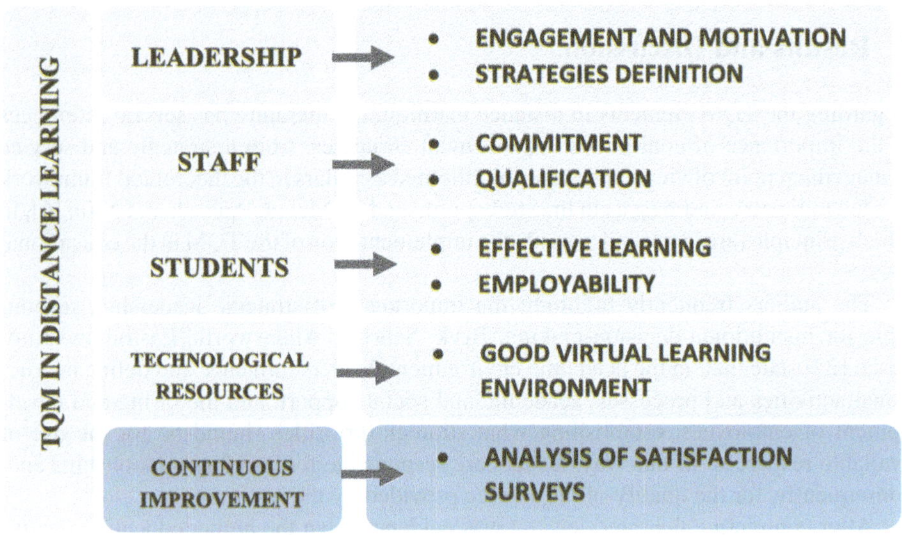

Fig. 1. Framework of TQM pillars in distance learning.

When dealing with the relationship with the customer in the educational environment, Gay and Betts [30] highlight the use of personalized communication strategies, which encourage regular and continuous interaction between instructor and students, whether through active participation in discussion forums or simply using Students' names respond to posts, creating a sense of closeness, in addition to providing personalized feedback on the assessed tasks. In the distance learning environment, technological resources are key elements for higher education institutions, according to Baig, Abrar, Ali and Ahmad [5], the educational world is closely connected to information technology, which allows students to access information and develop new skills, regardless of their location. Distance education becomes an effective way to gain leadership without the influence of social stereotypes and other barriers. Razik and Swanson [33] also emphasize that the use of electronic technologies enables the interaction of students and teachers separated in terms of geography, time, or both.

The focus on student satisfaction, with the maintenance of the relationship from enrollment to follow-up after completion of the course, appears as one of the main strategies to be applied to continue improving educational processes. An example of this is the award received by the institution object of this study due to the quality of the teaching material made available to students – a fact widely cited in interviews with managers. The use of augmented reality technology, support material with QR Code and virtual visits seeks to make studying more attractive for students, in addition to enabling their own management of study times and places. The bet on technological resources is a great differential for higher education institutions, since the mobile application allows the download of classes to attend at the time and place that is most convenient for each student.

Another essential point is the maximum use of available technology in favor of distance learning. For this, it is necessary to aim for resources that, in addition to being functional, are viable and accessible, as is the case of the mobile phone application, made available by the higher education institutions observed. The possibility for the student to have the study material at their disposal in an application, with the possibility of attending classes at the time that is most convenient for them, is not only a facilitator of learning, but also an interesting way to work on the issue of dropout. Many students opt for distance learning precisely because they work full-time and have little time available to attend classes at conventional times, in addition to the difficulty of daily travel to a university's headquarters. This is one of the main reasons for dropping out of the course, along with the financial issues involved. With the possibility of optimizing their study time, the academic gains another incentive to take the course forward.

Given these issues, it is understood that this study contributes to the literature on TQM by presenting a framework that associates the pillars defined in the literary findings with topics that stood out during observations and applied research, from the point of view of managers who experience daily the reality of higher education institutions focused on distance learning.

6 Conclusions

Since there is still a small number of articles that explore the TQM related to distance learning and, also, that the implementation of TQM in services is hampered by its particularities, such as heterogeneity, intangibility and perishability, the study guides future research to the development of the theme, as well as the application of the TQM in distance higher education institutions. This allows for a better-quality service, which is increasingly necessary, especially after the onset of the COVID-19 pandemic and the need to adapt to study and work.

From the identification of the practices adopted by the studied institution and its teaching centers associated with the practices identified in the literature, it is possible to answer the research question of this study. The resulting framework points to the way in which TQM principles can be incorporated in higher education institutions that only offer distance learning courses. The application of such guidelines enables the minimization or elimination of inefficiencies, the maintenance of competitiveness in the market, the agility in meeting customer demands, as well as providing a favorable environment for the work of employees and the learning of students, including cooperation between the same.

The elements presented in the proposed Framework can be seen as leverage points, capable of supporting the implementation of TQM principles in higher education institutions that offer distance learning. The result is a direction centered on the experience of educational processes, identified with the needs and interests of the stakeholders of these processes, indicating practices adopted with favorable effects for higher education institutions of different sizes and markets, which makes the delivery of this study even more interesting, as it is unrestrictedly applicable to any institution that seeks continuous improvement of its processes.

References

1. Alsoud, A.R., Harasis, A.A.: The impact of COVID-19 pandemic on student's e-learning experience in Jordan. J. Theor. Appl. Electron. Commer **16**, 1404–1414 (2021)
2. Sbaffi, L., Bennett, J.: Postgraduate students' experience of a jointly-taught, distance learning degree: the example of a Russell group university. J. Higher Educ. Policy Manag. **41**(6), 600–618 (2019)
3. Prakash, G.: Quality in higher education institutions: insights from the literature. TQM J. **30**(6), 732–748 (2018)
4. Nasim, K.: Role of internal and external organizational factors in TQM implementation: a systematic literature review and theoretical framework. Int. J. Q. Reliability Manag. **35**(5), 1014–1033 (2018)
5. Baig, S.A., Abrar, M., Ali, A., Ahmad, M.: Implementation of TQM on higher education in Pakistan (short communication). Qual. Quant. **49**(1), 51–56 (2013)
6. Mehta, N., Prakash, V., Nitin, S.: Total quality management implementation in engineering education in India: an interpretive structural modelling approach. Total Qual. Manag. Bus. Excell. **25**(1–2), 124–140 (2014)
7. Souza, M.A., Borchardt, M., Pereira, G.M., Viegas, C.V.: TQM at car dealerships with better sales performance: a multiple case study. Total Qual. Manag. Bus. Excell. **31**, 13–14 (2020)
8. Khan, M.A.: An empirical study of barriers in implementing total quality management in service organizations in Pakistan. Asian J. Bus. Manag. Stud. **2**(4), 155–161 (2011)
9. Martínez-Lorente, A.R., Dewhurst, F., Dale, B.G.: Total quality management: origins and evolution of the term. TQM Mag. **10**(5), 378–386 (1998)
10. Garvin, D.A.: Building a learning organization. Harv. Bus. Rev. **71**(4), 78–91 (1993)
11. Juran, J., Gryna, F.: Quality Control. Makron Books, São Paulo (1991)
12. Merli, G.: Eurochallenge: the TQM Approach to Capturing Global Markets. IFS, UK (1993)
13. Todorut, A.M.: The need of total quality management in higher education. In: 2nd world conference on educational technology researches (WCETR2012), Procedia - Social and Behavioral Sciences vol. 83, pp. 1105–1110 (2013)
14. Carvalho, M.M., Paladini, E.P.: Quality Management: Theory and Cases. Elsevier, Rio de Janeiro (2005)
15. Coltro, A.A.: Total quality management and its influences on business competitiveness. Caderno de Pesquisas em Administração 1(2), 106–107 (1996)
16. Paladini, E.P.: Quality Management: theory and practice, 2. ed. Atlas, São Paulo (2007)
17. Pedrotti, R.F.: Communication as a Quality Strategy in Organizational Management. Universidade Federal de Juiz de Fora, Juiz de Fora (2012)
18. Martínez-Caro, E., Cegarra-Navarro, J.G., Cepeda-Carrión, G.: An application of the performance-evaluation model for E-learning quality in higher education. Total Qual. Manag. Bus. Excell. **26**(5–6), 632–647 (2015)

19. Mello, C.H.P., Silva, C.E.S., Turriosi, J.B., Souza, L.G.M.: ISO 9001:2008: Quality Management System for Production Operations and Services. Atlas, São Paulo (2009)
20. Saraph, J.V., Benson, P.G., Schroeder, R.G.: An instrument for measuring the critical factors of quality management. Decis. Sci. **20**(4), 810–829 (1989)
21. Sordan, J.E., Marinho, C.A., Angeloni, M.: Assessment of perceived quality in a higher education institution: an action research. In: III National Production Engineering Symposium, vol. 3, no. 3, Universidade Federal da Grande Dourados, Dourados (2021)
22. Clemes, M.D., Cohen, D., Wang, Y.: Understanding Chinese university students' experiences: an empirical analysis. Asia Pac. J. Mark. Logist. **25**(3), 391–427 (2013)
23. Quinn, A., Lemay, G., Larsen, P., Johnson, D.M.: Service quality in higher education. Total Qual. Manag. Bus. Excell. **20**(2), 139–152 (2009)
24. Beketova, E., Leontyeva, I., Zubanova, S., Gryaznukhin, A., Movchun, V.: Creating an optimal environment for distance learning in higher education: discovering leadership issues. Palgrave Commun 6(1), 1–6 (2020)
25. Sallis, E.: Total Quality Management in Education, 3rd edn. Kogan Page, London (2002)
26. Tan, A.H.T., Muskat, B., Zehrer, A.: A systematic review of quality of student experience in higher education. Int. J. Qual. Serv. Sci. **8**(2), 209–228 (2016)
27. Bryk, A.S., Sebring, P.B., Allensworth, E., Easton, J.Q., Luppescu, S.: Organizing Schools for Improvement: Lessons from Chicago. University of Chicago Press, Chicago (2010)
28. Galeon, D.H., Garcia, P.G., dela Cruz, J.: E-learning roadmap for open distance learning in Cordillera Administrative Region. In: IOP Conference Series: Materials Science and Engineering 482 (1), 012012 (2019)
29. Garrison, D.R.: E-learning in the 21st Century: a Framework for Research and Practice. Routledge, Abingdon (2011)
30. Gay, G., Betts, K.: From distance forums to meetings: integrating high touch strategies to increase student engagement, academic performance, and retention in large online courses. Online Learn. J. **24**(1), 92–117 (2020)
31. Judge, T.A., Bono, J.E., Ilies, R., Gerhardt, M.W.: Personality and leadership: a qualitative and quantitative review. J. Appl. Psychol. **87**(4), 765–780 (2002)
32. Neborsky, E., et al.: Transition to distance learning under COVID-19 in assessments by professors. J. Phys. Conf. Series 1691(1), 100 (2021)
33. Razik, T.A., Swanson, A.D.: Fundamental Concepts of Educational Leadership and Management, 4th edn. Allyn e Bacon, Boston (2010)
34. Santally, M. I., Rajabalee, Y. B., Sungkur, R. K., Maudarbocus, M. I., Greller, W.: Enabling continuous improvement in online teaching and learning through e-learning capability and maturity assessment. Bus. Process Manag. J. **26,** 1687–1707 (2020)
35. Nadiri, H., Kandampully, J., Hussain, K.: Students' perceptions of service quality in higher education. Total Q. Manag. Bus. Excell. **20**(5), 523–535 (2009)
36. Bradbury, B.: A mixed methods study of distance learning graduate students' motivation and program satisfaction at an American State University. In: VII International Scientific and Technical Internet Conference, pp. 11–15. Belarusian National Technical University, Minsk, Belarus (2019)
37. Ribeiro, J. L.D., Milan, G.S.: Individual interviews: theory and applications. FEENG/UFGRS, Porto Alegre (2004)
38. Cavalcanti, L.M.R., Guerra, M.G.G.V., Gomes, C.S.F.: Evaluation of higher education courses in Brazil: the SINAES in its relationship with quality. Eccos Revista Científica **56**, 1–20 (2021)
39. Yin, R.K.: Case Study: Planning and Methods. 5th ed. Bookman Editora, Porto Alegre (2015)

Materials Selection in Product Development: Challenges and Quality Management Tools

Kristina Berladir[1], Justyna Trojanowska[2], Vitalii Ivanov[1(✉)], and Ivan Pavlenko[1]

[1] Sumy State University, 2, Rymskogo-Korsakova Street, Sumy 40007, Ukraine
ivanov@tmvi.sumdu.edu.ua
[2] Poznan University of Technology, Piotrowo Street 3, 61-138 Poznan, Poland

Abstract. Selecting the optimal material from a wide range of competitive variants significantly affects the final technical success of the product development. It avoids undesirable cost consequences and possible premature failure in the marketplace. Today, among the criteria for choosing materials, the emphasis has shifted more towards considering economic aspects and environmental impact. Various defects can occur during the life cycle of a material, from the production of original components to the commissioning of a product in the field. The nature of the deficiencies must be quickly identified and appropriate action taken. It has created a tendency to solve material selection problems using root cause analysis as the most useful problem-solving method in the engineering toolbox. The concepts and step-by-step methodology selected for research quality management tools in root cause analysis (Cause and Effect Analysis, Failure Mode and Effects Analysis, Quality Function Deployment, and Life-cycle Assessment) are considered in the paper. The main advantages and disadvantages of each method are established. Practical cases of using these methods to decide concrete applied problems in polymer materials science are described.

Keywords: Product innovation · Sustainable manufacturing · Root cause analysis · Cause and Effect Analysis · FMEA · QFD method · Life-Cycle Analysis · Life-Cycle Assessment · Process innovation

1 Introduction

At the base of materials science are the well-known paradigm (Fig. 1) about the relationship between the material structure, the processing methods for obtaining this material, and the resulting material properties. A complex combination of all these aspects provides material performance in a specific application [1].

Ultimately, materials affect product function, customer satisfaction, manufacturing systems, product lifecycle, who uses or manufactures it, usability, product identity, work environment, and cost [2].

Selecting an appropriate material or group for product design is often a complicated, responsible, and complex task for engineers. Material selection is a tedious task in

A. Hamrol et al. (Eds.): MANUFACTURING 2022, LNME, pp. 72–86, 2022.
https://doi.org/10.1007/978-3-031-00218-2_7

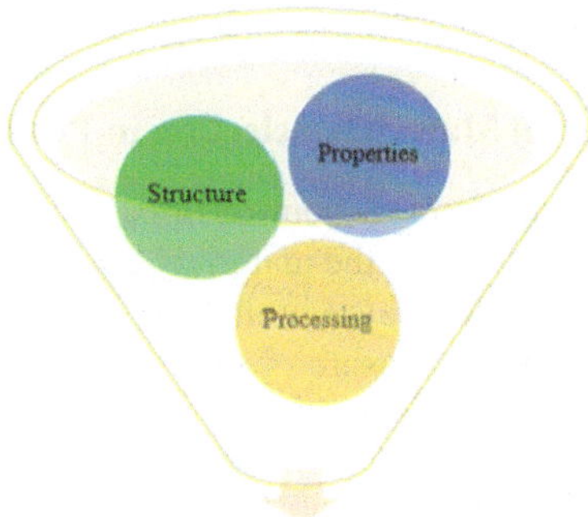

Fig. 1. Materials science paradigm.

manufacturing because many factors need to be carefully assessed before making a final decision [3]. The basic requirements may vary from application to application. Still, several other interrelated factors may also need to be considered depending on the operating environment, making the selection process more complex and time-consuming [4]. In the past, material selection was only a small part of the design process. Therefore, it did not receive the same level of research and development as other areas of design [2].

Nevertheless, few methods have been developed to solve material selection problems, some of which can be applied to material selection and any design selection procedure [5]. Today there is no one standard and universal method, but there are also widely used methods such as computer databases, performance indicators, decision matrices, expert systems, cost analysis, failure analysis, cost-benefit analysis, and so on [6].

Failure analysis is the logical and systematic evaluation of a component through analytical observation and verification of good engineering practice. Depending on the defect, the researcher may perform a damage analysis using macroscopic and microscopic examination, analysis of the composite material, assessment of the supramolecular structure, and physical or mechanical testing. Both indestructible and destructive testing can be performed. Although the selection of experimental materials, such as testing and prototyping, is the most accurate for a specific design solution, due to the time required and the high cost of experiments, especially if more material needs to be considered, this quickly becomes impossible [7].

Root cause analysis (RCA) is one of the easiest and simplest ways to solve a problem in an engineering toolbox. It is used to identify the root causes of failures (defects, malfunctions) in engineering systems and to eliminate them [8].

This paper presents the theoretical foundations of the concept of four management tools and practically shows how they can and should be used for visualization, assessment, and further analysis of polymer materials during operation.

2 Research Problem

2.1 Definition of Criteria on Materials Selection in Product Development

In practice, the material engineer must control three factors that affect the cost of a product: the design of the components, the materials, and the manufacturing technology (Fig. 2) [9]. These factors are interrelated in that the creation of an element can affect what material is used, and both the design of the component and the material used to impact the manufacturing technology choice.

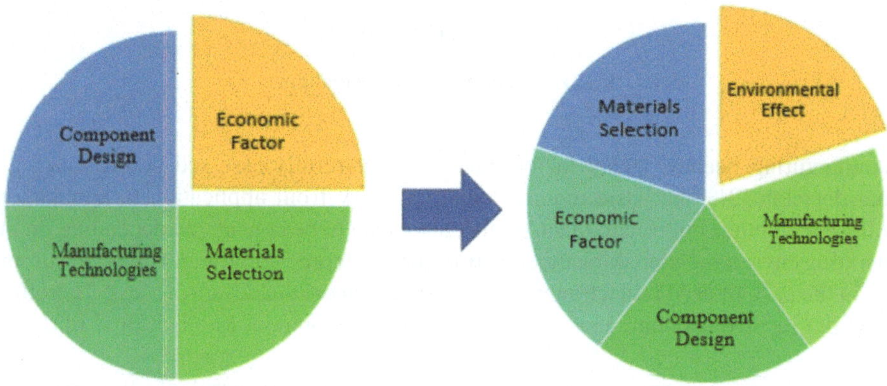

Fig. 2. Fundamental factors in sustainable manufacturing.

Some of the cost of a product component is related to its design (specification of size, shape, and configuration that affect the performance of the parts). Therefore, the design concept must consider the contribution of each component to the efficient operation of the entire system. The choice of materials with the appropriate combination of properties is based on the principle of affordability (least expensive). Therefore, a comparison is made of the cost of various competitive materials based on the price of each part. The choice of manufacturing process will be influenced by both the material selected and part design [10].

Materials play a crucial role in the "technology – economy – environment" scheme. So, the final product must be designed to have a minimal environmental impact. In addition, at the end of the service life, it must be possible to recycle the materials of the components, or at least dispose of them with minimal environmental impact [11].

2.2 Basic Methods of RCA for Development of Products or Services

Root cause analysis is used in lean production and helps to identify the root causes of the problem [12]. The main purpose of this method is to find the root cause of the problem through a series of questions: what happened, how it happened, and why it happened. This systematic approach to problem-solving is the root cause. It will help to develop solutions to the problem to prevent a recurrence. The general methodology of RCA is presented in Fig. 3.

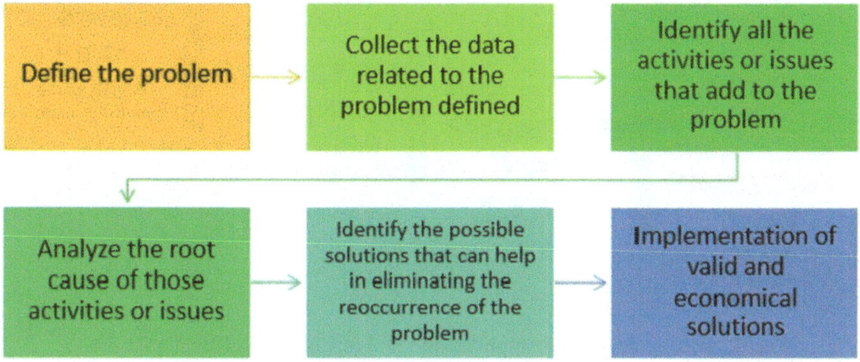

Fig. 3. Generalized RCA methodology.

RCA includes quite well-known and popular techniques for visualization, statistics collection, and quality analysis like 5 Whys, Fishbone diagram, FMEA method, Pareto method, etc. [13].

For our research, we settled on four of the most applied in the practice of materials science RCA tools: make a Fishbone diagram and analyze polymer composite defects by FMEA, make a house of the quality of polymer materials, and study the technique of LCA.

3 Results

3.1 Application of Cause and Effect Analysis for Defective Polymer Composite Material

The Cause and Effect Analysis helps identify the main factors that make the most significant contribution to the problem under consideration and prevent or eliminate their action. This method has already found its practical application in the diagnosis and defectoscopy of materials, in particular, to identify the leading causes of failure of the mounting bracket of the car bracket [14], chemical industry to assess the safety of spherical natural gas tank [15], biodiesel production technology [16], in automotive industries [17], etc.

The following steps illustrate the methodology of the Cause and Effect Analysis:

So, according to the main steps in Fig. 4, the materials science problem of obtained polymer composite blanks with low physical and mechanical properties was analyzed. These blanks are intended to manufacture various antifriction products by machining, including sliding bearings, O-rings of movable bearings and other friction units operating in air and liquid media, wet and dry gases in vacuum in the temperature range from minus 120 °C up to 260 °C. Blanks were obtained from a polymer composite, which does not meet the standard indicators (underestimated properties). The material of these blanks was Flubon 20 (80% polymer matric of PTFE and 20% carbon fibers) [18]. A Fishbone diagram was constructed (Fig. 5).

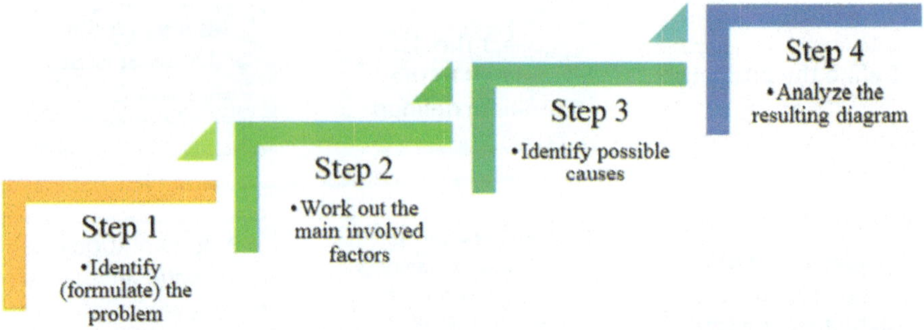

Fig. 4. Four main steps to applicate the tool of Cause and Effect analysis.

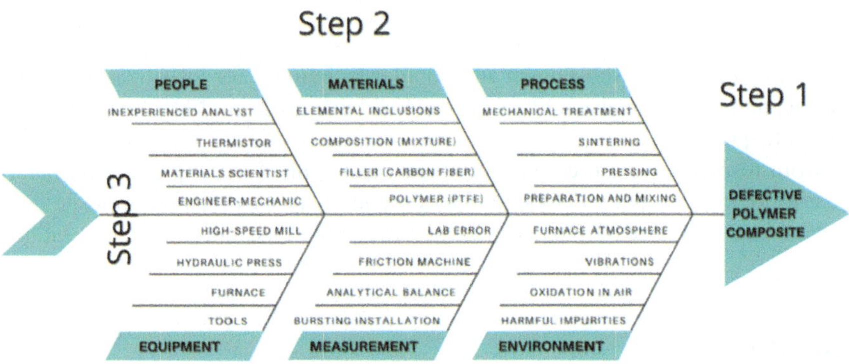

Fig. 5. Fishbone diagram of analyzing the defective polymer composite material.

A visual analysis of the machined turned surfaces of the test specimens revealed the following surface defects of the material surface: nonuniform distribution of carbon fiber in the polymer matrix, a patchy structure (alternating black and white spots), and the presence of cracks. During wear tests, an increase in temperature in the friction zone up to 200 °C was noted. It led to the seizure of the sample with the counter body. This phenomenon is associated with the material's low physical and mechanical properties (low density, uneven distribution of the filler in the polymer), confirmed by the tests carried out. All samples have a low elongation at break (30–100%). As a rule, this is a consequence of poor-quality mixing of the initial components – the presence of underground or coagulated fiber in the composition matrix.

The reasons were categorized into six key positions – person, method, material, mechanisms, control, and environment. The most significant causes of this problem were revealed with the employee's low qualifications and a violation of the mixing technology. In connection with the high responsibility of such products made from this material in the industry, it would be necessary to check the physical and mechanical properties of the entire batch. This problem would lead to the loss of material part, additional money and time, or replacement of the whole batch.

3.2 Investigation and Assessment of Macroscopic Polymer Composite Defects Using FMEA

At present, at least 80% of the processing of materials and technologies is carried out using the analysis of types and consequences of potential incompatibilities (FMEA methodology) [19]. The management of global companies widely introduces the types and consequences of potential incompatibilities for developing new designs and technologies, analysis, and planning of production processes and products [20]. This methodology makes it possible to assess the risks and potential damages from potential or existing defects from the initial design stage to create the finished product or its components [21].

In addition, the scope of the method covers all stages of the material life cycle, all technological processes, and trade [22]. The methodology of conducting any product/design FMEA involves following the steps (Fig. 6).

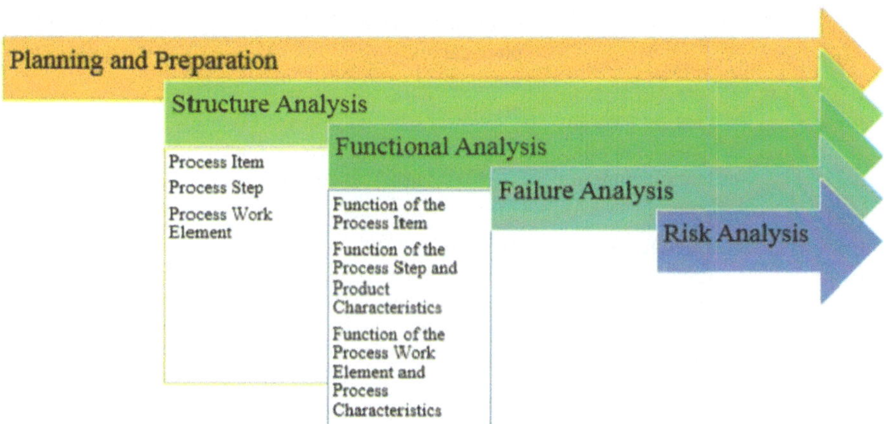

Fig. 6. The general methodology of the FMEA.

For example, FMEA was used to assess the macroscopic defects of polymer composite materials. External defects are various surface cracks, scratches, chips, chipping, etc.

At the macroscopic level, the following polymer composites characteristics were determined: uniform distribution of components; the number of components of the material, visually different in its structure; a geometric arrangement of the reinforcing filler; the geometric shape of the material components.

Finally, according to the FMEA methodology described in [23], it was assessed typically surface defects of polymer composites and established of permissible values for their characteristics. The results of the analysis are presented in Table 1.

Although surface defects slightly affect the mechanical properties of materials in the initial period of their occurrence, but they must be eliminated since the violation of the continuity of the outer surface of the material can further contribute to the penetration

Table 1. FMEA matrix.

Analyzed defect	Protentional causes	Possible consequences	S·O·D*, Risk assessment	Detection of the problem
Surface crack	Improper mode of forming the workpiece, high cooling rate. Shock effects during operation	Damage to the elements of the upper, inner, through detail part	9·3·3 = 81 defect is not critical (less than 100–120)	Determined visually, dependence on the size of the defect part, affected layers and direction of cracks
Scratch	Carelessness when removing parts from tooling and machining	Changes in the material composition in the cross section of the defect and physical and mechanical properties	5·1·3 = 15 are not large-scale	Determined visually, depth and width of the defect, orientation of the defect on detail
Inclusions	Getting foreign materials in the manufacture of prepreg		5·4·4 = 80 defect is not critical	Dimensions and thickness of the inclusion. Depth of occurrence
Chips, cuts and holes	Negligence when removing parts from equipment, transportation, storage, machining	They are voltage concentrators and reduce strength of composites	4·2·2 = 16 are not large-scale	Defect dimensions, width, depth
Local areas with high fiber content, matrix or pores	Non-compliance with the mode of formation, time and magnitude of the applied drop, heating rate	The degree of danger of these defects ultimately depends on the size, shape and location of the defect	9·4·3 = 91 defect is not critical	Determined visually, dependence on the size of the defect part, affected layers

* Notes: S (severity) – assessment of the most serious consequences of potential failure for consumers; O (occurrence) – the probability of specific causes and mechanisms of failure;D (detection) – assessment of the ability of the proposed management actions to detect this type of failure.

of aggressive media, which, ultimately, will significantly reduce the service life of the product.

Thus, this methodology can be successfully applied to all types of materials. It makes FMEA a universal management tool for assessing materials quality.

3.3 QFD Method

QFD is a flexible decision-making method that helps developers focus on the critical characteristics of a new or existing product or service from the perspective of an individual customer, market segment, company, or development technology [24]. The methodology results are clear diagrams and matrices that can be reused for future goods or services [25, 26].

In general terms, QFD consists of four matrices necessary to link the functions of the company and the needs of customers. These are product planning, part placement, process planning, and production planning schedules (Fig. 7). The initial QFD building process begins with a matrix that links customer needs to technical requirements and competitive information. This process is followed by a sequence of matrices that integrate the technical requirements into the design, operation and production system [27].

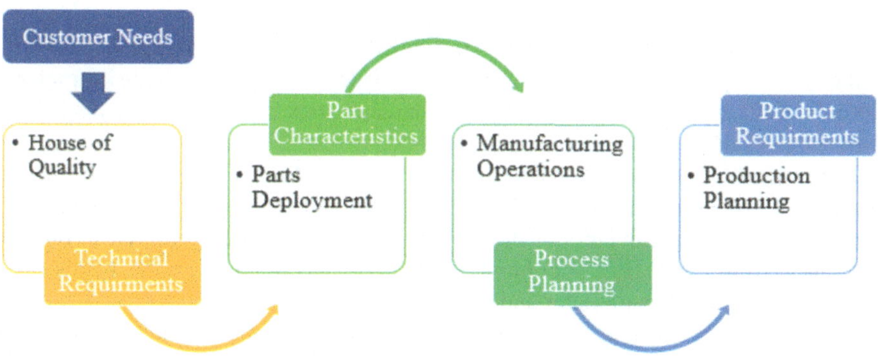

Fig. 7. Four QFD matrices.

The QFD quality function is deployed using a house-shaped diagram in which these matrices fit together (Fig. 1). It is named according to its form «House of Quality» (HOQ). The HOQ matrix is a basis that drives the entire QFD process and provides the tools to integrate customer requirements with technical requirements. The six building blocks of the HOQ matrix are described below (Fig. 8).

The central part of the house (3) is a table whose columns correspond to the technical characteristics (2), and the lines meet the requirements of the consumer (1). The level of addiction is noted in the cells, if any. The house's roof (5) represents information about the correlation between the technical characteristics. The left room (1) includes a column for the importance of customer requirements. The right room (4) consists of assessing the fulfillment of needs (from the consumer's point of view) for similar products on the market. The basement of the house (6) contains the results of the analysis of the technical

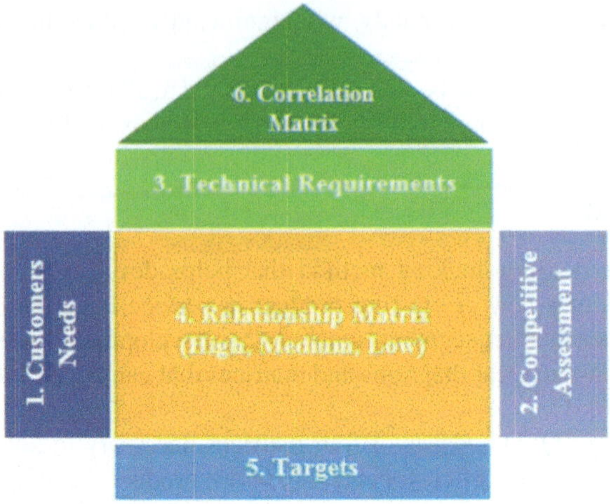

Fig. 8. House of quality matrix.

characteristics of competing products, the target values of the technical aspects of the products, estimates of the absolute and relative importance of the characteristics.

This methodology was used to carry out a quality analysis of developed polymer composites based on polytetrafluoroethylene with different fillers [28]. Figure 9 is shown for the most important indicators of products made of polymer composite material, where the links between requirements and characteristics are shown. The intensity of relationships is determined by the expert method.

The properties of products made of polymer materials are largely determined by the matrix's properties, the mixture's composition, and the modes of manufacturing processes. First, the parameters of the technological regime were ordered in order of importance (from 1 to 5) (Fig. 9, column 3). Then a connection was established between the technological mode parameter and characteristic (assign weights to strong, medium or weak links, respectively, 9, 3, 1). The transformation of customer requirements through QFD allows determining the relationship between requirements and characteristics. To determine the advantages of the developed materials in comparison with the industrial analogue, a comparison was made of the degree of consumer satisfaction (from 1 to 5).

Based on QFD analysis, it was found that the main parameters characterizing the required properties of the products under study are the strength at break, intensity of wear, and durability. The most significant factors affecting these parameters are material of polymer and filler and its technology of mixing. The influence of these parameters should be investigated to establish quantitative dependencies. The most priority technical characteristics were identified for further research and development of recommendations for improving quality.

3.4 LCA Analysis

Many of the industrial materials we consume come from non-renewable sources. These include most polymers that are used as raw materials for use, petroleum, and some

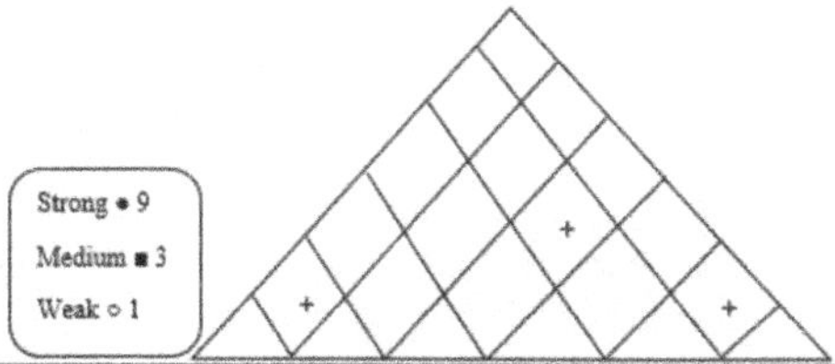

Row number	Requirement	Importance (1 to 5)	Characteristic: Strength at break	Relative elongation at break	Density	Intensity of wear	Coefficient of friction	Durability	Competitiveness assessment: Developed PTFE-composites [28]	Industry composite Flubon 20 [23]	Target value	Degree of improvement	Ponderability	Ponderability %
1	Type of polymer (matrix)	5	●	●	●	●	●	●	4	4	5	1,2	6,0	23,9
2	Type of filler	5	●	■	○	■	■	■	4	3	5	1,2	6,0	23,9
3	Mass ratio of polymer and filler	4	■	■	○	■	○	●	4	4	4	1,0	4,0	15,9
4	Mode of mechanical activation	3	●	●	○	●	○	■	3	0	4	1,3	3,9	15,5
5	The mixing time of the polymer with the filler	4	■	○	○	■	○	■	3	1	4	1,3	5,2	20,7
Absolute value			586,4	494,7	291,1	536,1	338,9	538,5						
Relative value, %			21	17,7	10,4	19,2	12,1	19,3						
Unit of measurement			MPa	%	kg/m^3	$mm^3/N{\cdot}m$	–	h.						

Fig. 9. House of the quality matrix for polymer composites.

metals. These valuable resources are gradually depleted, which requires new solutions: increasing resources, processing new materials comparable in volume to the environment, increasing results or developing new processing technologies. From the point of view of the economics of production, the influence of the environment and environmental factors on sustainable production, the life cycle of the material "from the cradle to the grave" is more important than the entire production process [10].

Life cycle assessment (LCA) includes assessing potential aspects of the environment and possible aspects related to material (or product). It is a systematic set of procedures for the collection and study of the import and export of materials and energy and their direct impact on the environment (ISO 14040:2006 «Environmental management. Life cycle assessment. Principles and framework») (Fig. 10).

The assessment of the product life-cycle is based on a rather complex and comprehensive methodology for assessing the impact of products on the environment at all stages of its life-cycle: extraction of raw materials, production, use, repair, maintenance, transportation of the product at all stages, and ending with disposal or recycling of waste. The main steps of LCA analysis are summarized in Fig. 11.

An excellent example of LCA was used to compare a car as an example of an individual vehicle and a bus as an example of public transport [29, 30]. The material flow consists of the impacts in the production, use and handling of related wastes. Data were collected on all material flows generated by the car and bus during their operation. It turned out that the tension of the material of the bus is much higher than that of the car, 37 times higher. However, given that a bus travels 7.4 times longer in its lifetime than a car and carries 25 times more passengers than a car, the impact of a public bus per kilometer and per passenger compared to a car is about 5.0 times less.

Based on a comparative review of the environmental impact of different types of traditional bioplastics [31], note that environmental aspects of plastics production are

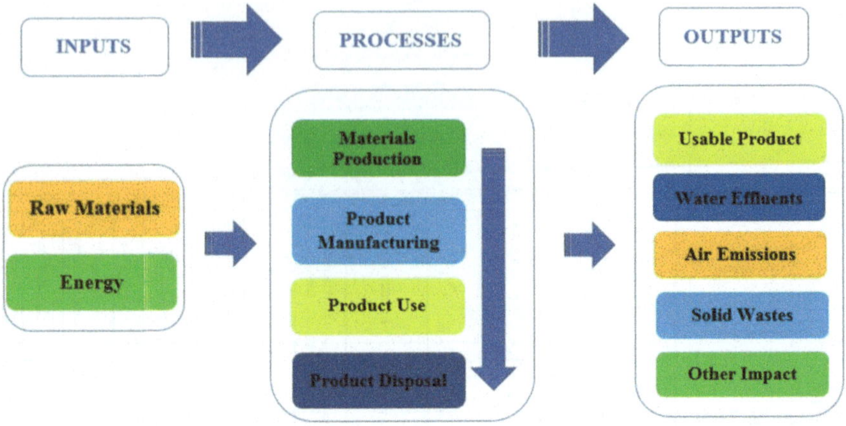

Fig. 10. Scheme of the life-cycle assessment of a product.

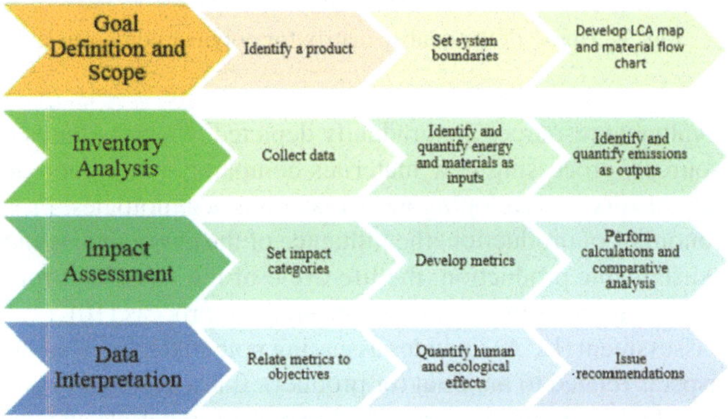

Fig. 11. Main key steps of a formal LCA study.

the negative impact on the atmosphere and protection of atmospheric air, as well as the formation of a few wastes in the production of plastic packaging. So, assessing the full life cycle developed polymer composites based on polytetrafluoroethylene with different types of fillers [28] seems to be an important and urgent task for future research direction.

4 Discussion

Cause and effect analysis visually display the relationship of the problem being investigated and the reasons that affect this problem; makes it possible to conduct meaningful research of the chain of interrelated causes affecting the problem; convenient and easy to use and understand.

The disadvantages of this quality tool include the difficulty of correctly determining the relationship between the problem under study. The causes of the problem under study are complex, i.e., part of a more difficult problem. The scope of analysis is quite broad, but it is often applied concerning developed products and related production processes.

The FMEA defines material properties that must be inspected, monitored, and controlled to ensure the part's reliability for the material is used to construct it. It is essential to understand that this type of analysis is intended to identify not the actual causes of probable failures but their manifestations, i.e., the types and forms of failures (failure modes).

The main advantage of QFD methodology over the other approaches is that it can be applied for processing qualitative and quantitative data and serve as a flexible framework that can be simplified, extended, and combined with other quality design and improvement techniques. Because this method provides both a score and a rank for each material, it helps designers better understand material selection issues by considering both the differences and similarities of alternative materials. Thus, it can be applied to any complex decision problem with many alternatives and quantitative criteria.

LCA of plastics and other structural materials play an important role in maintaining energy, saving resources, and saving money by preventing premature failures in an engineering component in a machine or equipment. LCA data on surface engineering materials improve the life cycle of a design component, increasing energy efficiency, sustainability and rejecting global temperature rise.

But LCA is applicable only as its data. Therefore, the data used to complete the life-cycle analysis must be accurate and up to date. When comparing different life-cycle analyses, equivalent data must be available for the product or process in question. If one product has much higher data availability, it may not be fair compared to another product with less detailed data. The importance of data is an ongoing challenge for life-cycle analysis. Thanks to globalization and the rapid pace of research and development, new materials and production methods are constantly being introduced to the market. It makes using up-to-date information very important and very difficult when performing LCA. If the LCA results are correct, the data must be up to date; however, the data collection process takes time. If the product and associated processes have not changed significantly since the last LCA data collection, the reliability of the data is not an issue. This paper will be useful for materials science and manufacturing engineers to understand the theoretical foundations of the concept of main management tools and practice using them with your own examples.

5 Conclusions

Cause and effect analysis turned out to be the easiest to understand and implement a method from RCA. To work with the Ishikawa diagram, we can easily involve students in practicing using management tools at the training stage.

QFD and FMEA, used as quality tools in composite materials design, can greatly improve development efficiency, leading to systematic design, material selection, and manufacturing.

The main LCA aim is to highlight the phases of the life cycle that have a strong impact on the environment and provide information for the product design process to improve environmental performance. This method is the only way to objectively determine which materials are environmentally friendly and which ones only seem to be.

Further research is related to the practical implementation of each of the described methods for solving various materials science problems of choosing materials. It is also planned to develop a prototype of the complex methodology (an expert system) based on multi-criteria decision-making (MCDM) system for materials selection in product development.

Acknowledgments. The authors acknowledge the research project "Fulfillment of tasks of the perspective plan of development of a scientific direction "Technical sciences" Sumy State University" (State Reg. No. 0121U112684), funded by the Ministry of Education and Science of Ukraine. The research was partially supported by the Research and Educational Center for Industrial Engineering (Sumy State University) and International Association for Technological Development and Innovations.

References

1. Johnson, N.S., et al.: Invited review: machine learning for materials developments in metals additive manufacturing. Addit. Manuf. **36**, 101641 (2020). https://doi.org/10.1016/j.addma.2020.101641
2. Jahan, A., Edwards, K. L., Bahraminasab, M.: Multi-criteria Decision Analysis for Supporting the Selection of Engineering Materials in Product Design, 2nd edn. Butterworth-Heinemann, Oxford (2016). https://doi.org/10.1016/B978-0-08-100536-1.00013-8
3. Wojciechowski, S., Przestacki, D., Chwalczuk, T.: The evaluation of surface integrity during machining of inconel 718 with various laser assistance strategies. MATEC Web of Conf. **136**, 01006 (2017)
4. Dobrotvorskiy, S., Balog, M., Basova, Y., Dobrovolska, L., Zinchenko, A.: Concept of the software for materials selection using .NET technologies. In: Tonkonogyi, V., et al. (eds.) InterPartner 2019. LNME, pp. 32–43. Springer, Cham (2020). https://doi.org/10.1007/978-3-030-40724-7_4
5. Rahim, A.A., Musa, S.N., Ramesh, S., Lim, M.K.: A systematic review on material selection methods. Proc. Inst. Mech. Eng., Part L: J. Mater. Des. Appl. **234**(7), 1032–1059 (2020). https://doi.org/10.1177/1464420720916765
6. Trojanowski P.: Comparative analysis of the impact of road infrastructure development on road safety–a case study. Scientific Journals of the Maritime University of Szczecin 23–28, (2020). https://doi.org/10.17402/436
7. Jansen, J.: Root Cause Analysis: Success in Failure. Plastics Engineering, http://read.nxtbook.com/wiley/plasticsengineering/julyaugust2014/rootcauseanalysissuccessinfail.html. Accessed 10 Oct 2021
8. Wojciechowski, S., Twardowski, P.: Tool life and process dynamics in high speed ball end milling of hardened steel. Proc. CIRP **1**, 289–294 (2012)
9. Ashby, M.: Materials Selection in Mechanical Design. 4th edn. Butterworth-Heinemann, Oxford. (2011). https://doi.org/10.1016/B978-1-85617-663-7.00024-2
10. Callister, W.D., Jr., Rethwisch, D.G.: Materials Science and Engineering: An Introduction, 10th edn. Wiley, USA (2018)
11. Djassemi, M.: A computer-aided approach to material selection and environmental auditing. J. Manuf. Technol. Manag. **23**(6), 704–716 (2012). https://doi.org/10.1108/17410381211253290
12. Mahto, D., Kumar, A.: Application of root cause analysis in improvement of product quality and productivity. J. Ind. Eng. Manag. (JIEM) **1**(2), 16–53 (2008). https://doi.org/10.3926/jiem.v1n2.p16-53

13. Doggett, A.M.: Root cause analysis: a framework for tool selection. Qual. Manag. J. **12**(4), 34–45 (2005). https://doi.org/10.1080/10686967.2005.11919269
14. Dondapati, S., Trivedi, M., Dondapati, R.S., Chandra, D.: Investigation on the mechanical stresses in a muffler mounting bracket using root cause failure analysis (RCFA), finite element analysis and experimental validation. Eng. Fail. Anal. **81**, 145–154 (2017). https://doi.org/10.1016/j.engfailanal.2017.08.010
15. Luo, T., Wu, C., Duan, L.: Fishbone diagram and risk matrix analysis method and its application in safety assessment of natural gas spherical tank. J. Clean. Prod. **174**, 296–304 (2018). https://doi.org/10.1016/j.jclepro.2017.10.334
16. Kaščak, J., et al.: Design of an atypical construction of equipment for additive manufacturing with a conceptual solution of a printhead intended for the use of recycled plastic materials. Appl. Sci. **11**, 2928 (2021). https://doi.org/10.3390/app11072928
17. Xu, Z., Dang, Y., Munro, P.: Knowledge-driven intelligent quality problem-solving system in the automotive industry. Adv. Eng. Inf. **38**, 441–457 (2018). https://doi.org/10.1016/j.aei.2018.08.013
18. Budnik, O.A., Sviderskii, V.A., Budnik, A.F., Berladir, K.V., Rudenko, P.V.: Composite material for chemical and petrochemical equipment friction assemblies. Chem. Pet. Eng. **52**(1), 63–68 (2016)
19. Sharma, K.D., Srivastava, S.: Failure mode and effect analysis (FMEA) implementation: a literature review. J. Adv. Res. Aeronaut. Space Sci. **5**(1&2), 1–17 (2018)
20. Petrescu, L., Cazacu, E., Petrescu, M.-C.: Failure mode and effect analysis in automotive industry: a case study. Sci. Bullet. Electric. Eng. Faculty **19**(2), 10–15 (2019). https://doi.org/10.1515/sbeef-2019-0014
21. Mascia, A., et al.: A failure mode and effect analysis (FMEA)-based approach for risk assessment of scientific processes in non-regulated research laboratories. Accred. Qual. Assur. **25**(5–6), 311–321 (2020). https://doi.org/10.1007/s00769-020-01441-9
22. Antosz, K., Pasko, L., Gola, A.: The use of artificial intelligence methods to assess the effectiveness of lean maintenance concept implementation in manufacturing enterprises. Appl. Sci. **10**, 7922 (2020). https://doi.org/10.3390/app10217922
23. Rudenko, S., Berladir, K., Trojanowska, J., Varenyk, S., Shvetsov, D., Kravets, V.: Application of FMEA for assessment of the polymer composite materials quality. J. Eng. Sci. **8**(2), B12–B18 (2021). https://doi.org/10.21272/jes.2021.8(2).b3
24. Kazemzadeh, R.B., Behzadian, M., Aghdasi, M., et al.: Integration of marketing research techniques into house of quality and product family design. Int. J. Adv. Manuf. Technol. **41**, 1019 (2009). https://doi.org/10.1007/s00170-008-1533-2
25. Prasad, K., Maity, S.R., Chakraborty, S.: Quality function deployment-based expert system for materials selection. Int. J. Mater. Sci. Eng. **4**(2), 80–93 (2016). https://doi.org/10.17706/ijmse.2016.4.2.80-93
26. Mayyas, A., et al.: Using quality function deployment and analytical hierarchy process for material selection of body-in-white. Mater. Des. **32**(5), 2771–2782 (2011). https://doi.org/10.1016/j.matdes.2011.01.001
27. Bhattacharyya, O., Chakraborty, S.: Q-analysis in materials selection. Decis. Sci. Lett. **4**(1), 51–62 (2015). https://doi.org/10.5267/J.DSL.2014.9.001
28. Berladir, K., Gusak, O., Demianenko, M., Zajac, J., Ruban, A.: Functional properties of ptfe-composites produced by mechanical activation. In: Ivanov, V., et al. (eds.) DSMIE 2019. LNME, pp. 391–401. Springer, Cham (2020). https://doi.org/10.1007/978-3-030-22365-6_39
29. Jacyna, M., Semenov, I.N., Trojanowski, P.: The research directions of increase effectiveness of the functioning of the RSA with regard to specialized transport. Arch. Transp. **35**(3), 27–39 (2015). https://doi.org/10.5604/08669546.1185181

30. Starzyńska, B., et al.: Requirements elicitation of passengers with reduced mobility for the design of high quality, accessible and inclusive public transport services. Manag. Product. Eng. Rev. **6**(3), 70–76 https://doi.org/10.1515/mper-2015-0028
31. Chernysh, Y., Balintova, M., Roi, I., Demcak, S.: Polymer materials: bibliographic data analysis and trend detection. Proc. Environ. Sci. Eng. Manag. **8**, 555–563 (2021)

Application of Random Forest Algorithm for the Quality Determination of Manufactured Surfaces

Marcin Hinz(✉), Jannis Pietruschka, and Stefan Bracke

Bergische Universität Wuppertal, Wuppertal, Germany
m.hinz@uni-wuppertal.de

Abstract. The optical perception of high precision, fine grinded surfaces is an important quality feature for these products. Its manufacturing process is rather complex and depends on a variety of process parameters (e.g. feed rate, cutting speed) which have a direct impact on the surface topography.

To improve the conventional methods of condition monitoring, a new image processing analysis approach is needed to get a faster and more cost-effective analysis of produced surfaces. For this reason, different optical techniques based on image analysis have been developed over the past years. Fine grinded surface images have been generated under constant boundary conditions in a test rig built up in a lab.

Within this study the image of each grinded surface is analyzed regarding its measured arithmetic average roughness value (Ra) by the use of random forest algorithms.

Keywords: Machine learning · Random forest · Computer vision · Condition monitoring

1 Introduction

The surface topography as well as the optical perception are important features for evaluating the quality of fine grinded knives. Parameters as the surface roughness, gloss or coloring are used for the quantification of these features. The measuring is implemented by the use of traditional methods, which are manual, time-consuming and cost-intensive. On top of that, the application of these methods for the condition monitoring of the ongoing process is rather limited. Therefore, a new, faster and more cost-effective approach is needed to improve the classical measurement methods. A conceivable approach could be based on image analysis.

Over the past years, different contactless image analysis based approaches have been developed to simplify the traditional roughness measurement methods. Some studies propose picture pre-processing and feature extraction in combination with machine learning algorithms. Neural Networks (NN) based methods are presented and tested on specific surfaces, as well (cf. [1–3]).

A. Hamrol et al. (Eds.): MANUFACTURING 2022, LNME, pp. 87–98, 2022.
https://doi.org/10.1007/978-3-031-00218-2_8

Fig. 1. An example image of a chef knife analyzed within this research activities

The overall goal of the presented research activities is the development of a condition monitoring tool which can be implemented in the ongoing grinding process of the knives. It should be used to ensure the knives quality and to reduce rejects by an immediate detection of deviations of the target values and the possibility to adapt the production process accordingly. For this reason, a data set based on cutlery samples has been generated and analyzed. During previous research activities, the extraction of features of the data set is presented. The features are used to train various machine learning algorithms with and without a combination of logged process parameters to evaluate the surface roughness (cf. [4, 5]). Another roughness prediction possibilities by the use of neural networks (NN) are presented in [6] and [7]. In this paper, the focus is set on the evaluation of the surface roughness (Ra) based on NNs. Therefore, a set of 851 chef's knives is used. An exemplary knife is shown in Fig. 1.

2 Data Generation

Within the presented research activities the analyzed data set embraces photographs of the surfaces of 851 8″ chef's knives (cf. Fig. 1). The surface images are taken within an experimental test rig which provides constant boundary conditions (cf. Sect. 2.2). Reference measurements of the surface roughness as well as gloss and coloring are taken of all the knives (cf. Sect. 2.1).

2.1 Reference Measurements

To describe the surface topography of the knife blades the surface roughness (Ra, Rq, Rz, Rt), gloss (GU) and coloring (CIELAB) are measured with traditional measurement methods. The roughness measurements are provided with the device PCE-RT11 over a sampling length of 6 mm. The arithmetic average roughness (Ra) can be calculated with Eq. (1). Here, z(x) is the surface profile and l describes the length of the measured section (cf. [4, 8, 9]).

$$\mathrm{Ra} = \frac{1}{l}\int_{0}^{l} |z(x)|dx \quad (1)$$

Since the Ra value is commonly used to define rejection limits in the regarded process, this parameter is used as the target value in this study. The measured Ra value of the data set varies between 0.07 µm and 0.47 µm.

2.2 Experimental Setup

In order to take comparable images of different knives, a test rig that provides consistent lighting conditions was designed and build up. The test rig is based on a design with white inside-walls, to keep ambient light outside and diffuse the light inside to prevent reflections on the knife's surface. The knives are mounted on a 3D printed fixation to ensure a proper positioning of each knife (Fig. 2).

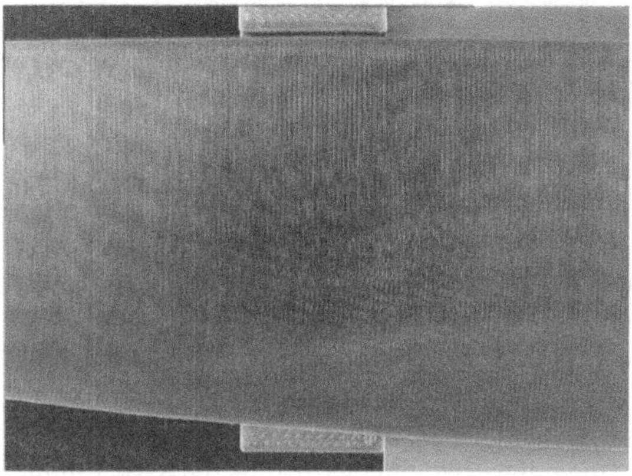

Fig. 2. An exemplary image of the photographed knife section

Two LED Spotlights, pointed at the walls on the top- and bottom ends of the knife, provide constant, diffuse light conditions inside the box. This light arrangement accentuates the characteristic marks left by the grinding process that appear in a 90° angle to the knife's length. The photos are taken by an Olympus E-520 DSLR, equipped with an Olympus Zuiko Digital 14–42 mm f 1:3.5–5.6 lens. The lens is placed normal to the knife's surface at a distance of 8 cm. Finally, to prevent reflections, a black screen at the height of 15 cm covers the camera body and the white ceiling, leaving only the lens visible from below. For a comprehensive description of the test rig cf. [4, 5].

3 Computer Vision Based Feature Extraction

Computer Vision (CV) describes the ability of perception of optical data by a computer. Since the investigation of optical data is extremely complex, picture pre-processing can be used to reduce the amount of information which will be studied to the most important characteristics for the available analysis task. The selection and application of appropriate pre-processing methods increase the quality and accuracy of the research. Therefore, within this study CV is used for analyzing the pictures of the knife surfaces and for the extraction of relevant features out of these images [1, 2, 10, 15].

As a first step the part of the knives, where the reference measurements had been taken, has to be identified on the pictures. In this way, scattering of the surface topography and inconsistent lightning conditions on the edge of the focused range don't affect the

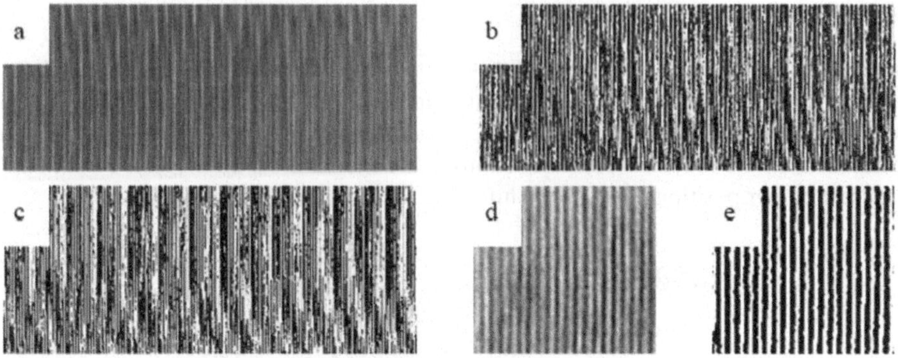

Fig. 3. (a) cropped image (b) sobel operator (c) contrast change and sobel operator (d) low pass mask and median blur (e) low pass mask, median blur and sobel operator

results of the further analysis. The images are cropped to a size of 1250 px × 550 px, which corresponds to an area of 2,5 cm × 2,0 cm.

The knife surfaces have a grooved structure, as it can be seen in Fig. 3(a). On all of the images the creases are rising vertically and parallel, but they differ in terms of the creases' width, depth and quantity along the considered picture section. As a result, information to determine these parameters will be detected within this research of the analysis of the surface roughness based on CV. Because the roughness is measured orthogonal to the creases, it makes sense to extract the features in the same way. The grooves are not consistent along their length. For this reason, ten uniformly distributed lines are drawn over the image height and the features are detected along [4, 8].

In order to get the best results, appropriate picture pre-processing filters and methods are utilized to reduce and adjust the kind of information which is inspected. The choices of the type of filters and their specification were made on base of researches and trials on the data base. For one part of the pre-processing the contrast of the images is changed. Besides that, low and high frequencies are filtered with approved filters. For the purpose of sharp separations between the individual grooves, the Sobel filter for the x-dimension with a kernel size of five is used. It is a well-known high pass edge detecting filter by which the kernel gets convolved with the image. As low pass filter, a filtering mask is used which is applied to the frequency domain of the image. On top of that, the median blur filter with a kernel size of three is selected to eliminate disruptive pixels by the comparison with its neighbors. Figure 3 shows the original and the pre-processed pictures [1–4, 11].

Subsequently, features are extracted. The creases' width and quantity are extracted from:

- the original image,
- the picture with the Sobel filter,
- the photo after the change of the contrast and the Sobel filter,
- the result after the application of the low pass mask and the median blur filter,
- the image with the use of the frequency mask, the median blur filter and the Sobel filter.

In addition to the absolute values, the averages and standard deviation of the creases' width of each line as well as their minimum, maximum and range are calculated. Between the results of each pre-processing combination obvious differences can be detected. Because the pictures are two-dimensional the grooves' depth cannot be gathered directly. Therefore, the lightness of each pixel is extracted by the use of the L*-value of the CIELAB color space. Since its course over the image width is comparable to a roughness profile, the course is used to determine parameters with the formulas of the roughness values (Ra, Rq, Rz, Rt). These parameters are calculated over the whole image width, over the width of the sampling length, and without the consideration of the edges of the images since these areas tend to show changing lightning conditions. That provides 42 features in total which are used within the following analysis:

- the number of creases,
- its average width and the associated standard deviation,
- the minimum and maximum width as well as the range of the crease's width from five differently pre-processed images (all above amount to 30 in total),
- roughness value formula calculated parameters (12 in total). cf. [5]

Exemplarily and according to [4], the average outputs of one line for two images can be found in the following Table 1.

Table 1. Results of the feature extraction on the basis of two exemplary picture for one line each

Filter	Image A		Image B	
	Average width of creases – line 1/pixels	Quantity of creases – line 1	Average width of creases – line 1/pixels	Quantity of creases – line 1
Original	4.34	287	3.70	335
Sobel operator	3.06	406	3.48	357
Contrast change	3.09	402	3.40	365
+ Sobel operator	14.77	31	14.32	37
DFT + median blur	13.82	39	12.71	42
DFT + median blur + Sobel operator				
R_a	0.11 µm		0.16 µm	
R_z	0.88 µm		1.56 µm	

It is obvious that all pre-processing stages result in different values. It is possible to detect less than 300 creases along one line from Image A out of the original picture. If the grooves are counted by human eye due to the verification purpose, the sum will be more than 400. That correlates with the result of the Sobel operator. Whether the contrast change is needed or not, depends on the individual photo. That is the reason why both results are used. The comparison of the groove width outlines the undetected edges of the original picture, as they are measured about one and a half time as wide in the original picture as in the filtered one. Apparently, the use of the discrete Fourier transformation causes the blur of several grooves into one. The identification works even better after edge detecting processes. By comparing the results of image A and B, in consideration of the references of their measured surface roughness, differences can be seen. On the one hand, it seems that there are more grooves on surfaces with a lower Ra value after the use of a high pass filter. On the other hand, the low pass filter causes less creases for lower surface roughness.

In general terms, regarding the results of this feature extraction, it can be stated that there are analogies between the measured roughness values and the extracted ones.

4 Random Forest Theory

In the previous subsection, data base and data generation were described. The machine learning model used for this study is the random forest which was first published in [12]. This Algorithm is a widely used forecasting tool and can be assigned to the learning technique of supervised learning. The application of the random forest relates to the classification and regression of data.

Random forests are a combination of tree predictors, also known as decision trees, such that each tree depends on the values of a random vector sampled independently and with the same distribution for all trees in the forest [12]. These types of algorithms generate classifiers or regressors in form of a decision trees by synthesizing a model based on a tree structure. The C4.5, one of the most known decision trees, first published in [13], is one of the most known and well renowned decision tree algorithms. Generally, is uses a divide-and-conquer method to grow an initial tree based on recursively partitioned data sets with the splitting criterion of entropy.

Entropy, or information entropy, is a representation of how much information is encoded by given data. Measured by the highest information gain, the data is split into partitioned data sets. At each node of a decision tree, the entropy belonging to a class membership is given by the formula:

$$info(S) = -\sum_{j=1}^{k} \frac{freq(C_j, S)}{|S|} \times log_2\left(\frac{freq(C_j, S)}{|S|}\right) bits \tag{2}$$

where freq(Cj, S) is the number of cases in S that belong to the class Cj. By applying formula 2 to a set of training classes, info(TT) measures the average amount of information needed to identify the class in the case of TT as follows:

$$info_x(TT) = \sum_{i=1}^{n} \frac{|TT_i|}{|TT|} \times info(TT_i) \tag{3}$$

And

$$gain(X) = info(TT) - info_x(TT) \tag{4}$$

with the outcomes of a test X. A disadvantage of decision trees is that they are prone to overfitting. As a result of overfitting the model becomes too catered towards the training data set and performs poorly on testing data. This will lead to low general predictive accuracy [14]. Whereas combining decision trees in a random forest circumvents this problem. To create the forests, the algorithm uses various methods of ensemble learning. The Bagging is a method for generating multiple versions of a predictor and using these to get an aggregated predictor. The aggregation averages over the versions when predicting a numerical outcome and does a plurality vote when predicting a class. The multiple versions are formed by making bootstrap replicates of the learning set and using these as new learning sets. This method adds some of embedded test data set to the model, which improves the accuracy on unknown data [12]. Besides bootstrapping, random forest using randomly selected subsets of features for creating slightly different tree structures. The simplest random forest with random features is formed by selecting randomly, at each node, a small group of input variables to split on. Consequently, the following properties are the result of the random forest improvements:

- It has been considered as very good algorithm in accuracy, also with noisy data
- It is much more efficient on huge data sets without overfitting
- Works fine without a lot of hyper parameter tuning
- Very good generalization ability
- Can process numerical data as well as strings

Another property of the algorithm should be mentioned separately. In the combination of trees, it is possible to see the importance of the individual features calculated by the algorithm. The Importance of the different features are measured by the mean decrease of impurity. The meaning of impurity in this context is the information content in relation to the forecast of the data. Therefore, the computed entropies of the features of all trees are summarized and averaged. With the insight into the importance of the features, it is possible to identify particularly relevant features in the data set. This method is used by for the reduction of input parameters presented in the following section.

5 Feature Reduction

Based on the mentioned impurities, the feature/input parameter reduction can be performed. For this purpose, the data set is split into two subsets consisting of production parameters (PP) and based on CV extracted features (EF). For the sake of simplicity, the parameters are numbered and don't need to be described in detail.

Once performing the feature weighting based on the impurities, as it can be observed in Fig. 4, we take a look into the subset based only on EF and without PP.

It can be observed that the feature 1 is the most important one with the highest mean decrease in impurity by around 8%. A slightly different behavior can be observed by applying the same method in a regression problem. In this case (not shown in a figure) the most important features are 1, 6 and 3 according to the mean decrease in impurity. The analysis of both, classification and regression problem in term of feature importance gives a diffuse projection of features. It can be stated that the first 6 features have slightly higher number than the remaining ones, though none of them (neither a single one nor a group) cab be defined as dominant.

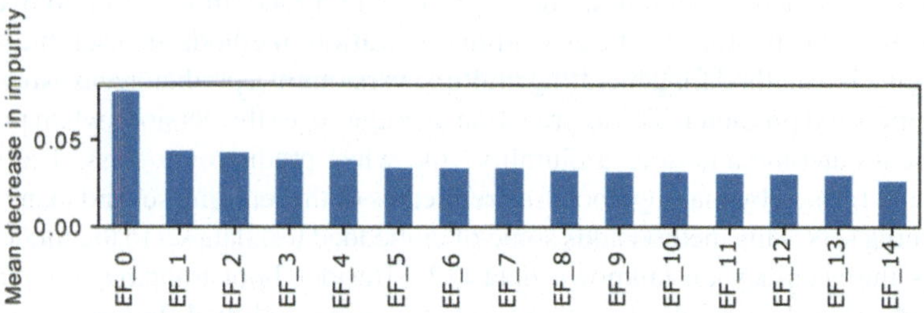

Fig. 4. Feature importance of classification with RF without PP

Once adding the PP into the data set, a different behavior can be observed (as shown in Fig. 5).

It is obvious in this case that the PP dominate the importance graph. The first extracted feature occurs at the amount of around 4% by a highest amount of the first production parameter (the temperature of cooling fluid) by around 13%. This parameter dominates even much more the importance in term of regression analysis with an amount of over 50%.

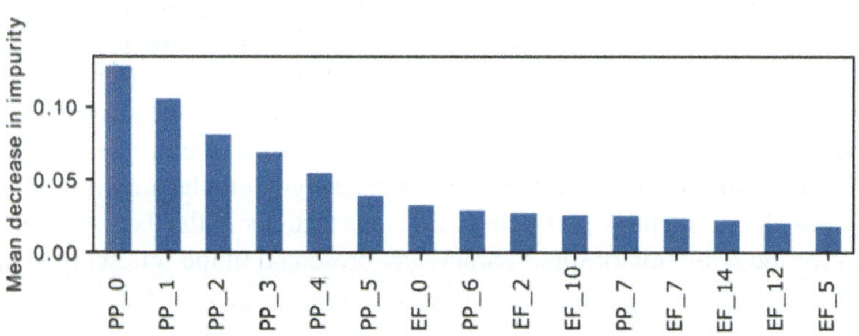

Fig. 5. Feature importance of classification with RF with PP

6 Analysis and Prognosis of the Data

In the following sections, a selection of numerical results of the machine learning analysis is discussed. Basically, all analysis processes were divided into many groups according to the certain problem. Therefore, we distinguish between different data sets (with and without PP) as well as between classification and regression problems, where the roughness values were classified by the specification limits into thee groups.

All the forests were trained with 80% of the data. The remaining subset was divided equally into two groups for the validation and test purposes. All the forests were performed with the amount of 1000 trees.

6.1 Prognosis Without Process Parameters

In Fig. 6, an exemplary confusion matrix obtained by the classification analysis of data without PP is shown.

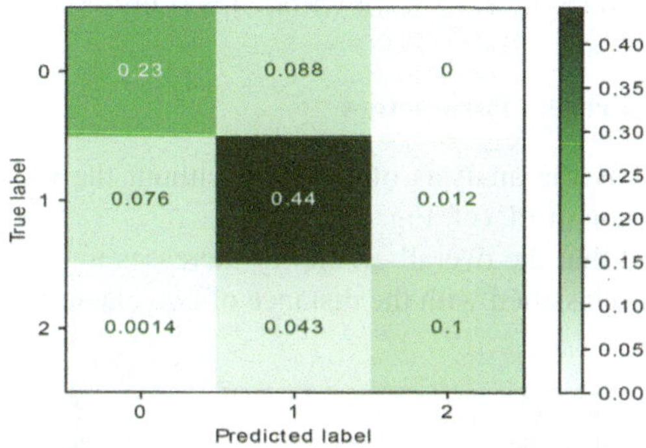

Fig. 6. Confusion matrix, classification problem, data set without PP

The overall efficiency of all perfumed analyses amounts to around 75%–77% which corresponds also to other tested algorithms that are not discussed within this paper (especially Support Vector Machines and Neural Networks). An important property of the analysis with RF is the fact that only very few data points are classified wrong with the distance of two classes (no wrong classified class 2 and very few in class 0). The most misclassification occurs especially at the borders of the specs which can let the conclusion of an inappropriate data amount.

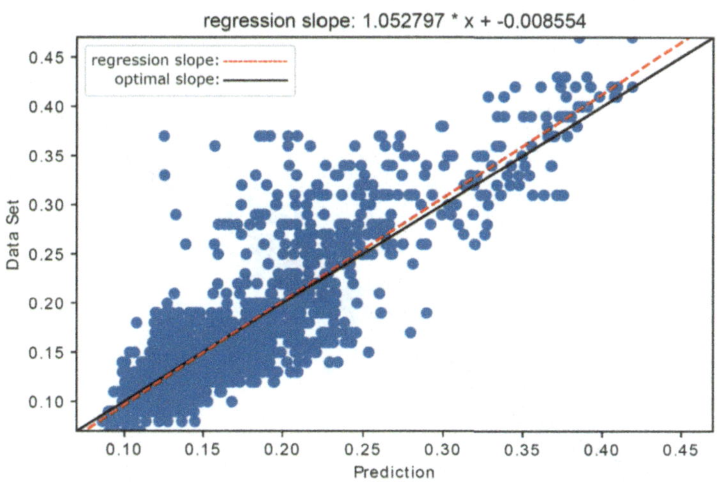

Fig. 7. Regression analysis of the data without PP – test phase

In Fig. 7 the results of the regression analysis in form of the predicted data as well as both models (predicted and real) in form of the two different lines are shown.

It shall be mentioned that the plot shows the test data which is separated only for the prediction purposes.

Once more it can be stated that the RF provide good results in terms of the model accuracy (mainly regarding the interpretation of both regression models but also based on quantified results which are not shown within this study).

6.2 Prognosis with Process Parameters

The achieved results by the analysis of the data without the PP can be significantly increased by the addition of PP (cf. Fig. 8).

It can be observed that the overall efficiency increases to 93% with the additional fact that no data is misclassified with the distance of two classes.

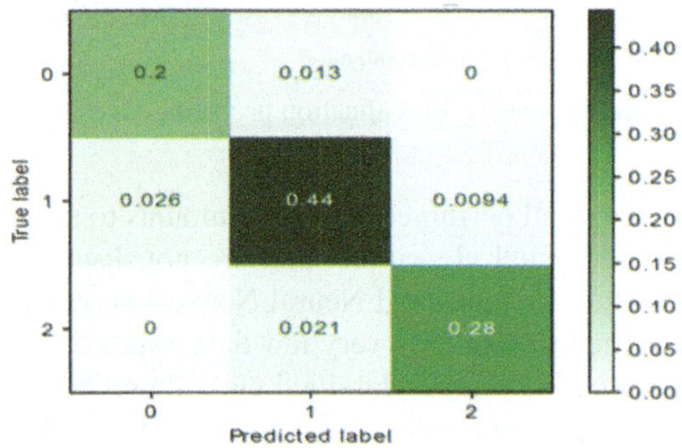

Fig. 8. Confusion matrix, classification problem, data set with PP

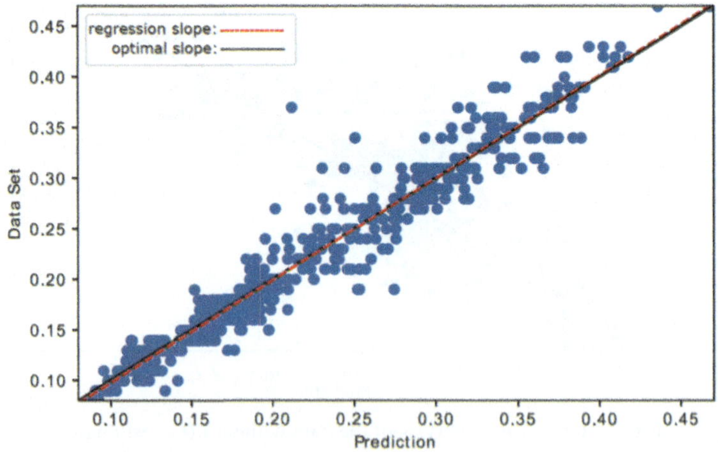

Fig. 9. Regression analysis of the data with PP – test phase

Also, the analysis of regression results leads to the conclusion that the addition of the PP increases significantly the modelling accuracy (as shown in Fig. 9).

Here, the scattering of the prediction decreases by a simultaneous increase of the model efficiency compared to the theoretical/optimal one.

7 Summary and Outlook

In this paper, the results of RF based analysis of fine grinded surfaces were shown. The training of the algorithms is based on feature extraction gathered by the means of CV. As summary, following declarations can be stated:

- RF provide comparable results to other machine learning algorithms which leads to the statement that the achieved efficiency of 75%–77% is mainly restricted by the feature extraction or even more particularly to the amount of unextracted information from the photos
- The inclusion of PP increases the efficiency of the algorithm to around 93%
- RF is a very good and well working algorithm for the feature reduction

In the further studies, more comprehensive feature extraction techniques shall be analyzed and performed. Here, a great number of potential features can be extracted and, with the help of FR, evaluated in terms of importance. A further possibility of the feature extraction is the application of LSTM [16] networks in terms of encoder – decoder algorithm. These networks with a recurrent part in the neurons itself give extract the features automatically, though not necessary with a higher technical understanding of the product.

Furthermore, an optimized camera system with an optimized lens within the test rig is planned to be tested. For an alternative measuring equipment in terms of the target variables, a confocal 3D measuring device shall be used.

References

1. Koblar, V., Pecar, M., Gantar, K., Tusar, T., Filipic, B.: Determining surface roughness of semifinished products using computer vision and machine learning. In: Proceedings of the 18th International Multiconference Information Society, Volume A, pp. 51–54 (2015)
2. Suen, V., et al.: Noncontact surface roughness estimation using 2D complex wavelet enhanced resnet for intelligent evaluation of milled metal surface quality. Appl. Sci. **8**, 381 (2018)
3. Rifai, A.P., Aoyama, H., Tho, N.H., Dawal, S.Z.M., Masruroh, N.A.: Evaluation of turned and milled surfaces roughness using convolutional neural network. Measurement **161**, 107860 (2020)
4. Hinz, M., Radetzky, M., Guenther, L.H., Fiur, P., Bracke, S.: Machine learning driven image analysis of fine grinded knife blade surface topographies. Procedia Manuf. **39**, 1817–1826 (2019)
5. Hinz, M., Guenther, L.H., Bracke, S.: Application of computer vision in the analysis and prediction of fine grinded surfaces. In: Baraldi, P., DiMaio, F., Zio, E. (eds.) Proceedings of ESREL 2020 PSAM 15. Research Publishing Services, Singapore (2020)

6. Pal, S.K., Chakraborty, D.: Surface roughness prediction in turning using artificial neural network. Neural Comput. Appl. **14**(4), 319–324 (2005)
7. Vasanth, X.A., Paul, P.S., Varadarajan, A.S.: A neural network model to predict surface roughness during turning of hardened ss410 steel. Int. J. Syst. Assur. Eng. Manag. **11**(3), 704–715 (2020)
8. DIN: DIN EN ISO 4287: 2010-07, geometrical product specifications (GPS) – surface texture: Profile method – terms, definitions and surface texture parameters (2010)
9. Bracke, S., Radetzky, M., Born, P.: Multivariate analyses of aperiodic surface topologies within high precision grinding processes. In: CIRP Conference on Intelligent Computation in Manufacturing Engineering (ICME). Gulf of Naples, Italy (2018)
10. Szeliski, R.: Computer Vision: Algorithms and Applications. Texts in Computer Science, Springer, London (2011). https://doi.org/10.1007/978-1-84882-935-0
11. Kekre, H., Gharge, S.: Image segmentation using extended edge operator for mammographic images. Int. J. Comput. Sci. Eng. **2**, 1086–1091 (2010)
12. Breiman, L.: Random forests. Mach. Learn. **45**, 5–32 (2001)
13. Quinlan, J.R.: C4.5: Programs for Machine Learning. The Morgan Kaufmann Series in Machine Learning, Kaufmann, San Mateo (1993)
14. Awad, M., Khanna, R.: Efficient Learning Machines: Theories, Concepts, and Applications for Engineers and System Designers, The Expert's Voice in Machine Learning. Apress Open, New York (2015)
15. Solem, J.: Programming computer vision with Python: [tools and algorithms for analyzing images]. Aufl. 1. ed. Beijing [u.a.]: O'Reilly (2012)
16. Yang, R., et al.: CNN-LSTM deep learning architecture for computer vision-based modal frequency detection. Mech. Syst. Signal Process. **144**, 106885 (2020)

Six Sigma and Random Forests Application for Product Quality System Control Development

Katarzyna Antosz[1(✉)], Arkadiusz Gola[2], Łukasz Paśko[1], Teresa Malheiro[3], Arminda Manuela Gonçalves[3], and Leonilde Varela[4]

[1] Faculty of Mechanical Engineering and Aeronautics, Rzeszow University of Technology, Al. Powstancow Warszawy 12, 35-959 Rzeszów, Poland
katarzyna.antosz@prz.edu.pl

[2] Department of Production Computerisation and Robotisation, Faculty of Mechanical Engineering, Lublin University of Technology, Lublin, Poland

[3] Research Centre, Guimarães, Portugal

[4] Algoritmi Research Centre, School of Engineering, University of Minho, Guimarães, Portugal

Abstract. Intelligent companies are able to become more and more advanced due to the accessible high-tech solutions, data collection and processing systems. Appropriate analysing and using of these large collections of data is significant in Industry 4.0 applications. The advancement of intelligent systems has a powerful effect on the progress of numerous innovative ideas considering the data analysis. They will make the choice of the proper data possible, but they will also suggest suitable methods for obtaining knowledge from these data. On that account, qualitative and quantitative research methods are employed in this paper to examine the outcomes on the product quality acquired from the case study company. The features which affect the product quality in a manufacturing process were recognized on the basis of statistical analysis of the acquired outcomes. Moreover, the work demonstrates that the machine learning method might be applied as a decision-supporting tool in a manufacturing process. The model created with the random forest method allowed to examine the influence of every single process parameter as well as the connection between them on the product quality. The acquired findings may be helpful while defining the most significant variables that might be considered to constitute the input data for developing an automated system for the uninterrupted monitoring of the process parameters, and, in consequence, for the product quality control.

Keywords: Product quality · Random forest · Six sigma project · Quality monitoring system

1 Introduction

Manufacturing processes are often characterized by abnormal changes of some key process parameters that could result in various categories of faulty products [1, 2]. Therefore,

A. Hamrol et al. (Eds.): MANUFACTURING 2022, LNME, pp. 99–112, 2022.
https://doi.org/10.1007/978-3-031-00218-2_9

a lot of provided research focuses on the problem of finding or developing the most appropriate model that would allow to identify the input parameters that are mainly responsible for the non-conformity in the specific production process. Weiss, Dhurandhar and Baseman described the methods for continual prediction of manufactured product-quality prior to final testing [3]. Guo, Li and Du proposed a model where an extreme learning machine is developed for monitoring the manufacturing processes and recognizing faulty quality categories of the products being produced [4]. An interesting two-step feature learning approach was introduced by Bai, Sun and Deng [5]. For the first step of learning, the multi-parameter feature is learned by a manifold learning algorithm (ML), while for the second step learning, the features of the low-dimensional information obtained by the ML are learned by a deep learning technique, which can learn sufficient features of the pattern between manufacturing quality and the low-dimensional information through layer-wise unsupervised training [6]. On the other hand, Ji presented the principle and the framework of quality prediction in intelligent manufacturing that adopt the theory of combinational prediction and allow to improve the accuracy of quality prediction as well as to reduce the production risk [7], while Schmitt and Deuse proposed an approach that utilizes quality prediction models and similarity algorithms [8].

In the last few years the provided research has mostly focused on developing methods and models based on data analytics and deep-learning techniques. In this field of study Jun, Chan and Jun presented a comprehensive framework that consists of three steps and allows to predict defects in improve yield by using the semi-supervised learning, time series analysis and classification model [9]. A deep multistage multi-task learning framework to jointly predict all output sensing variables in a unified end-to-end learning framework was presented by Yan et al. [10]. Recently, Zhang et al. has proposed a path enhanced bidirectional graph attention network (PGAT) that also allows to take into account dependency relationships among multiple machines in a multistage manufacturing process [11]. The comprehensive evaluation of several machine learning algorithms and the results of their performances in quality prediction was presented by Jung et al. [12].

Some of the initial recent research suggest the random forests technique as useful for the manufacturing quality prediction purposes. Although this technique is widely used in different scientific areas [13–16], its utilization for the manufacturing processes analysis and prediction is still very limited [17–20]. Therefore, the authors decided to develop the model which will help to analyse how each individual production process parameter and the relationship between them have an influence on the product quality. The data obtained from the case study company were used to develop the model.

2 Research Goal and Methodology Description

The main aim of the research was to develop a decision tool that can support monitoring of a production process in the company which produces paraffin inserts for candles. The main problem in the production process of this product is a large discrepancy of the product weight from the nominal value. Nonconforming products constitute approx. 5% of the production. Its most problematic consequence is the decrease in economic profits. If the weight tolerance is not sustained, the possibility of not providing a customer with the product specified in the earlier agreement grows. The research was performed to find the factors influencing the product quality. As a result, it could help the company in the process of making choices that would reduce the amount of nonconforming products. The results of the conducted research will be used as input data for the design of an automated system for continuous monitoring of the process parameters and, thus, for the product quality control.

The work methodology includes two phases. The first phase aims at recognizing the factors which are significant and which have an impact on the product quality statistically. The Six Sigma methodology was applied to explain the problem there. After that, the analysis of the production process was conducted according to the information obtained in the case study company. Then, the statistical analysis methods were used to examine the collected information. Additionally, in order to reinforce a decision-making process, machine learning methods (random forest) were applied in the second phase. The developed model presents the dependence of the production factors on the product quality.

3 Production Process Analyses with Six Sigma Methodology

3.1 Description of the Manufacturing Process

The first stage of paraffin inserts for the candle production process is the collection of raw materials from the warehouse. Then, the collected plastic containers are placed in a sorter. After filling the sorter, the containers are positioned. Then, they are transported to the machine responsible for sticking a wick together with the plate. This process is fully automated and it is called crimping. At the same time, on the second station, paraffin mass is prepared for filling the container. After the crimping stage, the blank goes to the production line and it is arranged in rows with the use of specialized combs which place the containers in the same distance from each other. Then, the containers are transported on the line where the first mass filling takes place. In this case, it is half of the target paraffin basis weight. The belt moves the containers towards the second filling machine at the appropriately set speed. During this time, the paraffin mass slowly solidifies. The next step is pouring the second layer with the paraffin mass. Next, the products move along a cooling tunnel where the mass solidifies completely. The next stage of the process is the product quality control. Products for the quality control are randomly selected. The quality control is carried out visually by the employees of the production line and the quality control department, as well as by photocells placed in a special tunnel. Identified defective products are withdrawn. The next step in the process is packaging. A stack of products is formed by means of a manipulator that grips aluminium containers with suction cups. The shaped stack is transported towards the foil wrapping machine. The foil-wrapped stack is labelled with a marking machine. The labelled package has a unique code including a production line, the last two digits of the year, the week of the year, the day, a production shift and time. The product marked this way facilitates the traceability of the products in the event of a complaint from a customer. The next stage is packing the stacks into a collective packaging, which is a flap carton, an easy-to-open display with skidding, a display for ½ pallets or stacks placed loosely on a pallet. During the packing process, a visual inspection of the wrapped packages takes place. Then, the goods are properly secured, wrapped with stretch foil and taken to the ready products warehouse. In Fig. 1 the general overview of the analysed process is presented.

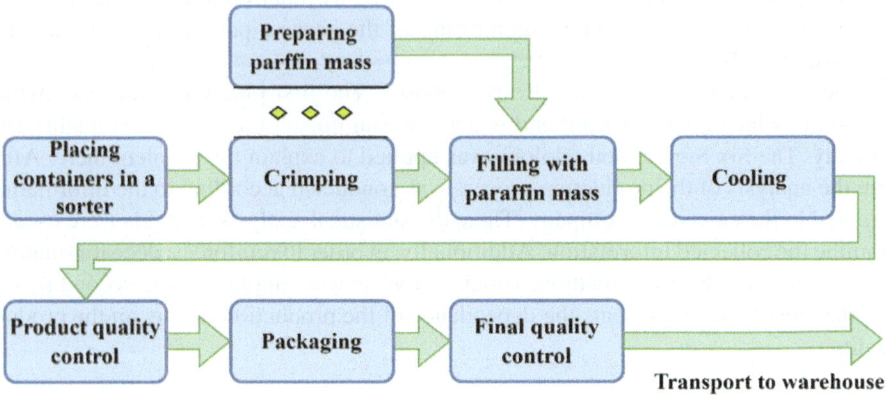

Fig. 1. The general overview of the analysed process.

3.2 Six Sigma Project - Problem Definition, Goal and Scope

The problem of incorrect weight of the product, which causes high costs of the production processes in a company, was a subject of the Six Sigma (SS) project. Till now, the case study company have tried to minimize the number of nonconforming products, for example implementing the quality management tools and lean manufacturing methods. However, the client's requirements incline the company to decrease the nonconforming products. In order to solve this problem, the company started to implement the SS project. The aim of the SS project was to identify the factors which have influence on the quality of the product (product weight within specification). The implementation of the SS project will reduce the production costs resulting from the production of non-conforming products and stabilize the process on a production line.

3.3 SIPOC Development, Data Collection and Analyzes

In the first phase of the SS project, in order to present the production process, a SIPOC diagram was developed. The SIPOC diagram includes: material suppliers for the process (S), Inputs (x1 - temperature of the paraffin mass in the tank), main process operations (P): wick crimping, paraffin mass pouring, cooling, packing product for different customers (C). The following process data were collected: x1 – paraffin mass temperature in the tank [°C], x2 - worm temperature [°C], x3 - baths temperature [°C], x4 - filling machines temperature [°C], x5 - nozzle temperature [°C] x6 - priming time [s], x7 - temperature in the tunnel [°C], x8 - cooling time [s], x9 - day of the week, x10 - shift, x11 – operator. The data were collected from 1 March, 2021 to 30 June, 2021. In this period the sample with 740 observations was taken from the process. The gathered data were the subject of a further analysis. In Table 1 the basic statistics for input variables are presented.

Table 1. Basic statistics for input variables.

Variable	Average value	Min.	Max.	Standard deviation
x1	41,40	40,60	44,20	0,809
x2	21,54	20,10	22,50	0,760
x3	56,17	55,00	57,00	0,617
x4	42,78	40,00	45,30	1,225
x6	1,02	1,00	1,20	0,052
x7	15,92	15,30	16,50	0,335
x9	2,96	1,00	5,00	1,454
x10	2,02	1,00	3,00	0,824
x11	2,02	1,00	3,00	0,824

For the identification of the factors that have the influence on the product quality and corelation between them, due to the type of the analyzed data, the statistical *chi* - square and Spearman tests were performed. The Spearman's rank correlation coefficient (R) was determined by an error according to the formula (1):

$$r_s = 1 - \frac{6\Sigma_{i=1}^{n} d_i^2}{n\left(n^2 - 1\right)} \tag{1}$$

where: d_i^2 - squares of differences between the ranks of the corresponding feature values x^i and y^i; n – a number of data pairs (number of rows in the table). The *chi*-square statistic was calculated from the following formula (2):

$$\chi^2 = \sum \frac{(O - E)^2}{E} = \sum_{i=1}^{k}\sum_{j=1}^{p} \frac{\left(n_{ij} - E_{ij}\right)^2}{E_{ij}} \tag{2}$$

This statistic has an asymptotic distribution $\chi^2 with (k-1)(p-1)$ degrees of freedom. Where: O – observed values; n_{ij}; E_{ij}– theoretical values obtained from the product of border values. The *chi*-square test is often used to test the independence of two qualitative variables. In this case, a two-way table was created, and the expected frequencies for the cell were determined as the product of the total count in the column, and the total count in the row divided by the total number of observations. For the depended variable (product quality – OK/NOK) the *chi*-square test with null hypothesis $H_0\colon x_1 = x_2 = \cdots = x_{11}, and H_1\colon x_1 \neq x_2 \neq \cdots \neq x_{11}$, with confidence level $\alpha = 0.05$ was performed.

Tables 2 and 3 present the results of statistical tests which were performed to test the hypotheses.

The results presented in Table 2 were analyzed according to the Cohen scale [21]. According to this scale there is no correlation if the value of R varies around 0.1, the

Table 2. Results of Spearman and *chi*-square test for inputs variables.

Variable	x1	x2	x3	x4	x5	x6	x7	x9	x10	x11
x1	-	0.273	-0.243	-0.029	-0.029	-0.095	0.322	-0.161	-0.008	-0.008
p-value	-	**0.000**	**0.000**	0.427	0.427	0.010	**0.000**	**0.000**	0.830	0.830
x2	0.273	-	0.112	-0.323	-0.323	0.388	0.499	-0.126	-0.081	-0.081
p-value	**0.000**	-	**0.002**	**0.000**	**0.000**	**0.000**	**0.000**	**0.001**	**0.027**	**0.027**
x3	-0.243	0.112	-	-0.015	-0.015	0.394	-0.226	0.036	-0.081	-0.081
p-value	**0.000**	**0.002**	-	0.677	0.677	**0.000**	**0.000**	0.332	**0.027**	**0.027**
x4	-0.029	-0.323	-0.015	-	1.000	-0.351	-0.259	-0.042	0.079	0.079
p-value	0.427	**0.000**	0.677	-	0.011	**0.000**	**0.000**	0.259	**0.031**	**0.031**
x5	-0.029	-0.323	-0.015	1.000	-	-0.351	-0.259	-0.042	0.079	0.079
p-value	0.427	**0.000**	0.677	0.011	-	**0.000**	**0.000**	0.259	**0.031**	**0.031**
x6	-0.095	0.388	0.394	-0.351	-0.351	-	0.332	-0.035	-0.062	-0.062
p-value	**0.010**	**0.000**	**0.000**	**0.000**	**0.000**	-	**0.000**	0.338	0.091	0.091
x7	0.322	0.499	-0.226	-0.259	-0.259	0.332	-	-0.162	-0.033	-0.033
p-value	**0.000**	**0.000**	**0.000**	**0.000**	**0.000**	**0.000**	-	**0.000**	0.367	0.367
x9	-0.161	-0.126	0.036	-0.042	-0.042	-0.035	-0.162	-	0.013	0.013
p-value	**0.000**	**0.001**	0.332	0.259	0.259	0.338	**0.000**	-	0.725	0.725
x10	-0.008	-0.081	-0.081	0.079	0.079	-0.062	-0.033	0.013	-	1.000
p-value	0.830	**0.027**	**0.027**	**0.031**	**0.031**	0.091	0.367	0.725	-	0.011
x11	-0.008	-0.081	-0.081	0.079	0.079	-0.062	-0.033	0.013	1.000	-
p-value	0.830	**0.027**	**0.027**	**0.031**	**0.031**	0.091	0.367	0.725	0.011	-

correlation is weak if R varies around 0.3, medium if R varies around 0.5 and large if R is more than 0.5. For most of the analyzed variables there is no correlation or correlation is weak. For some of them (highlight in green) the correlation is medium. These are, for example, variables x4 and x2. The highest correlation is between variables x2 and x7 ($R = 0.499$ – highlighted in blue). Although the correlation between the factors was identified, a cause-effect relationship might not exist. That is why the analysis showed a relationship between the variables for which p–value was < 0.05). Table 3 presents the results of statistical tests for inputs and product quality (Q).

On the basis of the statistical tests results presented in Table 3, it can be said that the factors which have influence on the products quality are: paraffin mass temperature in the tank [°C] (x1; p-value = 0.000), worm temperature [°C] (x2; p-value = 0.000556), baths temperature [°C] (x3; p-value = 0.000008) and the temperature in the tunnel [°C] (x7, p-value = 0.000). Moreover, the medium correlation between x1 and product quality ($R = -0.473$) can be observed.

The result of the analyses indicated the factors that have a great impact on the product quality from the statistical point of view. Moreover, it also revealed the correlation both between the individual parameters of the process and the product quality.

Therefore, the question arises which of the analyzed factors should be taken into account in the first place by the company to build a system of continuous monitoring of the product quality control. Which of them are the most important? In order to answer

Table 3. Results of statistical tests for inputs and product quality (Q).

Variables x_j & Q	N	R	t(N−2)	p-value
x1	740	−0.473	−14.587	**0.000**
x2	740	−0.127	−3.467	**0.005**
x3	740	0.163	4.495	**0.008**
x4	740	0.072	1.961	0.050
x5	740	0.072	1.961	0.050
x6	740	0.004	0.1091	0.913
x7	740	−0.210	−5.825	**0.000**
x9	740	0.027	0.735	0.462
x10	740	0.037	1.013	0.311
x11	740	0.037	1.013	0.311

this question, the authors used the machine learning (random forest) method to develop a model in a further stage of the study, which explains the changes in the product quality depending on the process parameters. In addition, this model will allow to identify the importance of the variables that have the greatest impact on the quality of products.

4 Random Forests to Support the Decision-Making

4.1 Random Forests

When it comes to scientific methods, companies can find data mining methods helpful. There are several data mining procedures originating from well-founded scientific areas such as statistics and machine learning. The random forests technique is among them. It is a means of classification (and regression) that includes forming a number of decision trees established on a random collection of data. This algorithm aims at creating a board of experts from random decision trees. On the contrary to the classic decision trees, the rule underlying the random trees is that a subset of the analysed features in the node is chosen in a random way. The random forest method is perfect for sample testing with a high observation vector. The learned random forest may be used not only for classification but also for different tasks, and this is considered an extra value. The Random Forest [22] consists of a set of classifiers (ensemble classifier), where every single classifier stands for a decision tree learned in an uninterrupted manner (till merely the samples of the identical class remain in the leaf). Every single classifier of the random forest is trained on a D0 data sample, created particularly for it, and which is formed by drawing n times with returning from all N learning samples [23]. Furthermore, in the process of creating a decision tree, at the point of defining the decision rule in a node some attributes are ignored. M attributes are outlined, where $m < M$. The decision rule in the node is specified according to them. When it comes to predicting a class for a sample, it is shown to all decision trees, the final choice considering the class affiliation is established on the outcome selected by most decision trees in the random forest. The random forest algorithm has many decision trees and which creation may take place in an independent way. Constructing a single decision tree in the random forest involves looking thoroughly for all the obtainable variables as well as all probable splits in a dataset for every decision node by means of choosing an optimal split [20]. In order to construct a random forest, it is necessary to observe the features (dependent and independent variables) and create a dataset $D = \{(y_i, x_i)\}_{1 \leq i \leq n}$. For each observation the dependent (output) variable y_i may assume values from a set $\{c_1, c_2, \ldots, c_s\}$, although e.g. the vector of realizations of independent variables (predictors) goes into R^k space, $x_i = (x_{i1}, x_{i2}, \ldots, x_{ik}) \in R^k$. All the achievable classes of the variable y are symbolised by the values $c_1, c_2, \ldots, c_s$. The classification problem mainly involves predicting a suitable class the feature will belong to established on the acquired information (observation vector) of the independent variables $x_i = (x_{i1}, x_{i2}, \ldots, x_{ik})$.

Effectiveness of solution of the presented task depends on dividing R^k space on q separated regions, where each region corresponds to an appropriate class. To create classification tree the entire space R^k is divided into q separated regions, $R_1 \cup R_2 \cup \cdots \cup R_q = R^k$. For each region is calculated the Gini index value, also called the node pollution measure (or impurity measure). For the node m, $1 \leq m \leq q$, representing region R_m, the Gini index is determined as follows [24]:

$$Q_G(m) = \sum\nolimits_{j=1}^{s} p_{mi}(1 - p_{mi}) = 1 - \sum\nolimits_{j=1}^{s} p_{mi}^2 \tag{3}$$

where p_{mi} is a conditional probability for $i-$ th class in $m-$ th node, s – a number of classes. For node m with n_m observations the conditional probability for $j-$ th class is equal:

$$p_{mi} = \frac{\#\{y = c_i : x \in R_m\}}{n_m} \tag{4}$$

The fundamental purpose of the random forests decreasing a number of variance by diminishing the correlations between the trees. In order to reduce the variance one needs to grow the tree sets, in which the predictors for every tree are selected in a random manner, i.e. in the first place j predictors, $j \leq k$, are chosen this way, then R^j space is split in order to construct a tree. Each tree's output variable classification (predicted class) from the random forest (a set of trees), founded on the observation of the vector of predictors$(x_{i1}, x_{i2}, \ldots, x_{ik})$, is obtained through a majority vote strategy.

When it comes to the random forest model, it provides a higher level of stability and classification outcomes over a decision tree algorithm. However, it is comparatively more complex considering the interpretability of classifiers. Furthermore, random forests give the opportunity to classify the tree variables from the forest in order to determine their predictive accuracy. The random forest algorithm constructs a tree form an original dataset D after the set $D_0 \subset D$ is pulled at random. Samples that are non-random are termed as out-of-bag (OOB) [23].

The OOB set helps to assess prediction errors as well as the importance of single variables. The OOB prediction error expresses a number of possible test set samples which have not been properly classified. That's the difference between the elements positioned in the validity matrix and the elements positioned outside the diagonal of the matrix. Several different parameters are applied to evaluate the classification model [25–28]. The errors by class are indicated in a confusion matrix. They are described as: the first type errors when the objects are classified from the positive to negative class, and the other error type in case of assigning the objects from the negative class to the positive one [29]. The potential resulting classifier outcomes are specified as: TP (true positive) – refers to the correctly indicated positive samples, TN (true negative) - correctly indicated negative cases, FP (false positive) – refers to the samples wrongly indicated as positive, FN (false negative) - the samples wrongly indicated as negative. On the basis of the confusion matrix, the evaluation of the predictive value of a random forest may be established with the help of diverse measuring instruments. However, the model accuracy is the most significant.

4.2 Model Development

A random forest was built for the developed research problem. The primary aim involved creating:

- a model which would show how the product quality changes dependent on the values of the considered parameters,
- the significance standing of the variables together with the data on the parameters which influence the product quality at most.

The collected output information was assigned into the categories: the product quality, OK (good quality) and NOK (nonconforming) products. Therefore, the investigated task is considered a classification problem.

While developing the random forest two parameters of the model building algorithm were altered, namely the tree number in the model and the number of parameters pulled into the pool. The issues considering the parameters which influence the product quality, the single parameters as well as the relations between them affecting the product weight were recognised.

The algorithm that built the random forest combined the following criteria: the sort of the random forest classification, the number of trees and m-try (i.e. the number of predictors to be tested at random at each split during the creation of the tree models). Furthermore, the dataset including 740 observations was distributed into two sets. Namely, these are: teaching (aimed at learning the classifier and consisting of 80% of the instances) and validating (designed for the classifier trial and consisting of the remaining 20% of instances). The percentage ratio 70 to 20 was made purposefully in order to enhance the probability of achieving better effects.

4.3 Results and Discussion

The classification model of the random forest was primarily generated for the following data: ntree = 200 and mtry = 4 (assuming that mtry = sqrt (p) principle, where p = 11 of analysed input features). The above parameters used in a training dataset served to design a random tree with the resulting classification errors for all OOB variables at 1.08%, in the case of the OK class = 0.0039 and for the NOK class = 0.0561. The minimization of the classification error designed for the OOB set and for OK and NOK classes depends on the trees number (Fig. 2).

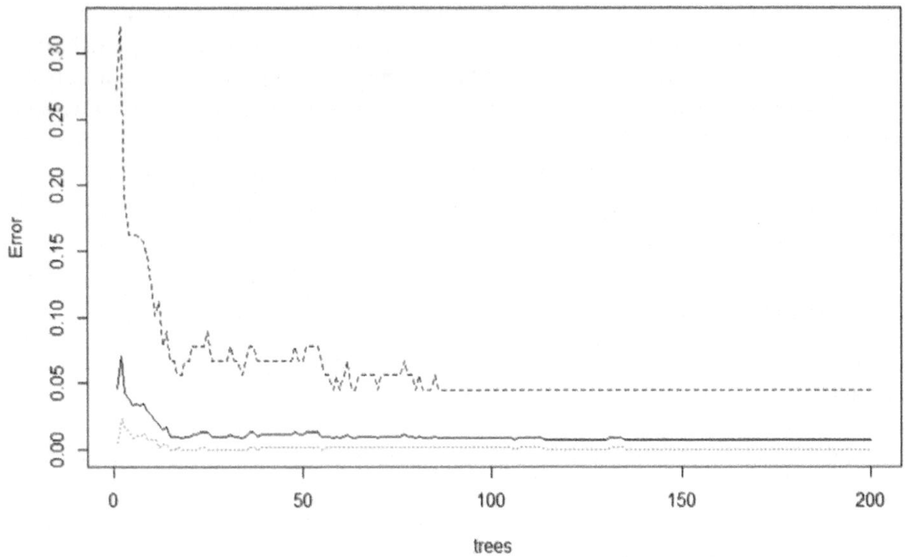

Fig. 2. The observed change in the OOB estimate and the classification error for the OK and NOK classes for the ntree value of 200.

Figure 2 demonstrates the change in the OOB error, and the classification error the OK and NOK classes for ntree = 200. A rapid decline of the error can be observed in the beginning. However, starting with the number of 80 trees, the increase of the base models ceases to reduce it in a considerable way.

The constructed model predictive potential was tested for the training and validation data with parameters of ntree = 100 and mtry = 4. The confusion matrix for the training data was presented in Table 4. The attention should be drawn to the fact that the classification error of OOB equals 0.68%. Four instances were casted to the incorrect classes, thus the prediction Accuracy is 0.993 (4/592). The prediction error was acquired on the level of 0.67% ((1-Accuracy) * 100%).

Table 4. Training data - confusion matrix.

	Reference		
	Quality	NOK	OK
Prediction	NOK	86	3
	OK	1	502

The model was also checked for the validation data set. Table 5 shows the confusion matrix for the validation data.

Table 5. Validation data - confusion matrix.

	Reference		
	Quality	NOK	OK
Prediction	NOK	25	1
	OK	1	121

In a validation dataset two items were wrongly classified. It results in the accuracy rate of 0.9864. The great value and effectiveness of the designed model is proved by a low prediction error at 1.36% level.

One of the most significant characteristics of the established models of the random forest is their potential to define the importance of particular variables. Their importance is computed for every single random forest that was designed. It is offered in a system of variable weights. The variables ranking is based on the mean variations in the Gini index. It provides the way to prioritize the variables according to their importance. Therefore, it establishes how important the variables are for the successful classification, thus which have greater predictive attributes. The higher the weight is, the bigger effect it has on the described variable. The importance ranking for the most significant variables of the constructed model is shown in Fig. 3. The highest scores mean the greatest impact of the variable on the product quality.

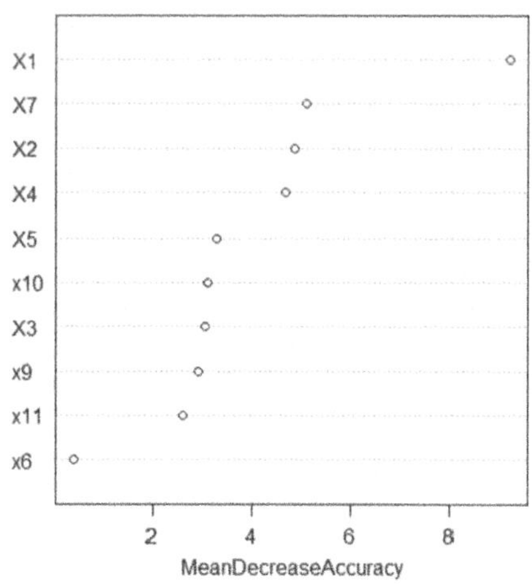

Fig. 3. Significance ranking for the most important variables of the created model.

The importance of the variable determines the participation of the variable in the created random forest. Importantly, the specific meaning of a variable applies only to the analysed for which it was determined. Interpreting the above values, it can be concluded that the best variable in terms of predictive values among the variables analyzed is variable x1 (paraffin mass temperature in the tank [°C]), x7 (temperature in the tunnel

[°C]), x2 (worm temperature [°C]) and x4 (filling machines temperature [°C]). The lowest variable is x6 (priming time [s]). Variable x1 is particularly important compared to other variables. The results of analysing the variable importance for a random forest can be used to determine the most important input variables, while rejecting those which do not affect the quality of the products. Moreover, three of the identified most important variables in the model are compatible with the results obtained after the performed statistical analyses. These are the variables: x1, x2, and x7. These variables should be included in the first place in the monitoring system of the products quality of paraffin inserts for candles.

5 Conclusions

The work aimed primarily at showing a method which facilitates the identification of the elements that have an impact on the product quality. The offered approach is the combination of machine learning method and the Six Sigma methodology. At first, the Six Sigma method was applied to examine the issue. Next, the identification of possible factors likely to affect the product quality was accomplished through statistical analyses. Based on their results, the following factors which have an effect on the product quality were recognised: paraffin mass temperature in the tank, warm temperature baths temperature and the temperature in the tunnel.

In addition, a decision making process was facilitated in the second phase by the use of the random forest algorithm. What is more, the constructed model aided the determination of the most significant factors as well as their relation to the product quality. The key variables were established on the basis of the factors analyses performed for the random forest model. They were as follows: paraffin mass temperature in the tank, the temperature in the tunnel, worm temperature and filling machines temperature. Three of the variables defined as of the greatest importance in the model overlap with the outcomes of the carried out statistical analyses. The factors described as the most significant ones ought to be integral in the process of constructing the product quality monitoring system.

In spite of several advantages of the offered approach, there are some constraints as well. The study was performed in a single company and the methodology hasn't been applied in the real conditions. The authors' further study will consider the application of the offered approach in constructing the product quality monitoring system. Moreover, the assessment of the usefulness of this approach as well as its potential for customising to different conditions will examined based on the achieved outcomes and their analysis.

References

1. Ivchenko, O., et al.: Method for an effective selection of tools and cutting conditions during precise turning of non-alloy quality steel C45. Materials **15**(2), 505 (2022). https://doi.org/10.3390/ma15020505
2. Kotowska, J., Markowski, M., Burduk, A.: Optimization of the supply of components for mass production with the use of the ant colony algorithm. In: Burduk, A., Mazurkiewicz, D. (eds.) Intelligent Systems in Production Engineering and Maintenance – ISPEM 2017, pp. 347–357. Springer, Cham (2018). https://doi.org/10.1007/978-3-319-64465-3_34

3. Weiss, S.M., Dhurandhar, A., Baseman, R.J.: Improving quality control by early prediction of manufacturing outcomes. In: 19th ACM KIGKDD International Conference on Knowledge Discovery and Data Mining (KDD 2013), pp. 1258–1266 (2013)
4. Guo, J., Li, Y.B., Bai Gang, D.: Dynamic quality prediction of manufacturing process based on extreme learning machine. Adv. Mater. Res. **889–890**, 1231–1235 (2014)
5. Bai, Y., Sun, Z.Z., Deng, J.: Manufacturing quality prediction based on two-step feature learning approach. In: International Conference on Sensing, Diagnostics, Prognostics and Control (SDPC), pp. 260–263 (2017)
6. Deng, J., Bai, Y., Li, C.: A deep regression model with low-dimensional feature extraction for multi parameter manufacturing quality prediction. Appl. Sci. **10**(7), 2522 (2020)
7. Ji, M.: Research on the framework of quality prediction in intelligent manufacturing. In: Wang, K., Wang, Y., Strandhagen, J., Yu, T. (eds.) Advanced Manufacturing and Automation VII, vol. 451, pp. 573–579. Springer, Singapore (2018). https://doi.org/10.1007/978-981-10-5768-7_60
8. Schmitt, J., Deuse, J.: Similarity-search and prediction based process parameter adaptation for quality improvement in interlinked manufacturing processes. In: IEEE International Conference on Industrial Engineering and Engineering Management (IEEE IEEM), pp. 700–704 (2018)
9. Jun, J.H., Chang, T.W., Jun, S.: Quality prediction and yield improvement in process manufacturing based on data analytics. Processes **8**(8), 1068 (2020)
10. Yan, H., Sergin, N.D., Brenneman, W.A., Lange, S.J., Ba, S.: Deep multistage multi-task learning form quality prediction of multistage manufacturing systems. J. Qual. Technol. **53**(5), 526–544 (2021)
11. Zhang, D.H., Liu, Z.Y., Jia, W.Q., Liu, H., Tan, J.R.: Path enhanced bidirectional graph attention network for quality prediction in multistage Manufacturing process. IEEE Trans. Industr. Inf. **18**(2), 1018–1027 (2022)
12. Jung, H., Jeon, J., Choi, D., Park, J.Y.: Application of machine learning techniques in injection molding quality prediction: implications on sustainable manufacturing industry. Sustainability **13**(8), 4120 (2021)
13. Devotta, S., Chelani, A., Vonsild, A.: Prediction of flammability classifications of refrigerants by artificial neural network and random forest model. Int. J. Refrig. **131**, 947–955 (2021)
14. De Freitas, A.G.M., Minho, L.A.C., de Magalhaes B.E.A., Dos Santos, W.N.L., Santos, L.S., Fernandes, S.A.D.: Infracted spectroscopy combined with random forest to deterimine tylosin residues in powdered milk. Food Chem. **365**, 130477 (2021)
15. Lim, D.K., Mustapha, K.B., Pagwiwoko, C.P.: Delamination detection in composite plates using random forests. Compos. Struct. **278**, 114676 (2021)
16. Burduk, A., Musiał, K.: Genetic algorithm adoption to transport task optimization. In: Graña, M., López-Guede, J.M., Etxaniz, O., Herrero, Á., Quintián, H., Corchado, E. (eds.) International Joint Conference SOCO'16-CISIS'16-ICEUTE'16. Advances in Intelligent Systems and Computing, vol. 527, pp. 366–375. Springer, Cham (2017). https://doi.org/10.1007/978-3-319-47364-2_35
17. Zhou, H., Yu, K.M., Chen, Y.C., Hsu, H.P.: A hybrid feature selection method RFSTL for manufacturing quality prediction based on a high dimensional imbalanced dataset. IEEE Access **9**, 29719–29735 (2021)
18. Arrais-Castro, A., Varela, M.L.R., Putnik, G.D., Ribeiro, R.A., Machado, J., Ferreira, L.: Collaborative framework for virtual organisation synthesis based on a dynamic multi-criteria decision model. Int. J. Comput. Integr. Manuf. **31**(9), 857–868 (2018)
19. Kunz, G., Machado, J., Perondi, E., Vyatkin, V.A.: Formal methodology for accomplishing IEC 61850 real-time communication requirements. IEEE Trans. Industr. Electron. **64**(8), 6582–6590 (2017)

20. Cohen, S., Kamarck, T., Mermelstein, R.: Perceived stress scale. Measuring Stress Guide Health Soc. Sci. **10**(2), 1–2 (1994)
21. Araújo, A.F., Varela, M.L.R., Gomes, M.S., Barreto, R.C.C., Trojanowska, J.: Development of an intelligent and automated system for lean industrial production, adding maximum productivity and efficiency in the production process. In: Hamrol, A., Ciszak, O., Legutko, S., Jurczyk, M. (eds.) Advances in Manufacturing. LNME, pp. 131–140. Springer, Cham (2018). https://doi.org/10.1007/978-3-319-68619-6_13
22. Breiman, L.: Bagging predictors. Mach. Learn. **26**, 123–140 (1996)
23. Breiman, L.: Random forests. Mach. Learn. **45**, 5–32 (2011)
24. Antosz, K., Jasiulewicz-Kaczmarek, M., Paśko, Ł., Zhang, C., Wang, S.: Application of machine learning and rough set theory in lean maintenance decision support system development. Eksploatacja i Niezawodnosc Maintenance Reliab. **23**(4), 695–708 (2021). https://doi.org/10.17531/ein.2021.4.12
25. Costa, E.P., Lorena, A.C., Carvalho, A.C.P.L.F., Freitas, A.A.: A review of performance evaluation measures for hierarchical classifiers. In: Evaluation Methods for Machine Learning II: Papers from the AAAI-2007 Workshop, pp. 182–196. AAAI Press (2007)
26. Fawcett, T.: An introduction to ROC analysis. Pattern Recogn. Lett. **27**, 861–874 (2006)
27. Powers, D.: Evaluation: from precision, recall and F-score to ROC, unforcedness, nakedness & correlation. J. Mach. Learn. Technol. **2**, 37–63 (2011)
28. Provost, F., Fawcett, T., Kohavi, R.: The case against accuracy estimation for comparing classifiers. In: Proceedings of the ICML 1998, pp. 445–453. Morgan Kaufmann, San Francisco (1998)
29. Pasko, Ł., Setlak, G.: Examination of quality of predictive market segmentation. Zeszyty Naukowe Politechniki Śląskiej, Seria Informatyka, 37, **1**(123), 83–97 (2016). (in Polish)

Evaluating the Statistical Process Control Data Acquisition System in a Heat Exchanger Factory

Diogo Costa[1(✉)], Vaibhav Shah[2,4], Leonilde Varela[1,4], Caetano Monteiro[3,4], Goran Putnik[1,4], and José Machado[3,5]

[1] Department of Production and Systems, Universidade do Minho, Guimarães, Portugal
dioewerton@gmail.com
[2] Department of Information Systems, Universidade do Minho, Guimarães, Portugal
[3] Department of Mechanical Engineering, Universidade do Minho, Guimarães, Portugal
[4] ALGORITMI Research Center, Universidade do Minho, Guimarães, Portugal
[5] MEtRICs Research Center, Universidade do Minho, Guimarães, Portugal

Abstract. Manufacturers of Heating Ventilation and Air Conditioning systems in Brazil are experiencing several challenges to increase competitiveness of their operations, considering the world class companies which are continuously improving their products and processes to deliver better quality products with lower costs. To achieve this target, new opportunities for improvements in management and operation system are welcome. This paper evaluates the statistical process control data acquisition system in a company in Manaus Industrial Zone for heat exchanger manufacturing, by using BPMN for Process Mapping at current state. Following the mission of Industry 4.0, options for improvements were proposed for a new future state, considering quality improvements and resources reduction. The analysis brought process improvements, for instance increased speed of data analysis time, regarding the corrective and preventive actions taken in the production system, and its influences in reducing the failure rate of products due to leakage rate, besides the analysis of the failure rate of product performance, and the reduction of resources used to collect process data.

Keywords: SPC · Data acquisition · BPMN · Industry 4.0

1 Introduction, Context and Methodology

Brazilian industry has been facing a challenging period of high international competitiveness, including attraction of other countries with lower production costs [1]. They are also experiencing substantial changes with Industry 4.0 [2]. The world has gone through several stages of industrialization and revolutions occur with each paradigm shift. We are going through the 4th Industrial Revolution, which brings several impacts on industrial sector, starting with new business models development consumer demand increasing, and the industrial model should undergo an evolution in order to serve the new markets and quality and reliability of products improvements [3].

Additionally, unexpected events can cause more adverse scenarios in economics, like outbreak of COVID-19, which results in a recession effect, demand decrease, and

A. Hamrol et al. (Eds.): MANUFACTURING 2022, LNME, pp. 113–124, 2022.
https://doi.org/10.1007/978-3-031-00218-2_10

disruption in supply chain, and a possible recovery time is impossible to predict and it will depend on many factors for economic activities restart [4], what causes new initiatives for process improvements to be pointed and evaluated.

Considering the actual scenario described above, new alternatives for quality improvements, cost reduction and production increase are requested to these companies to be more competitive. Study as [5] shows a bibliometric study of papers in last 50 years considering SPC-related themes, [6] shows an example of application of BPMN for monitoring and improving a Maintenance System and [7–9] show importance of Lean Six Sigma integration with BPMN (Business Process Model Notation).

This paper presents a case study of split Heating, Ventilation Air Conditioning (HVAC) components assembly and scenarios for improvement. It will be shown research to consider refrigeration systems and components fabrication, for promoting the company to the transformative requisites underlying the Industry 4.0 era. Then current state is described by using a BPMN map constructed. Improvement scenarios are constructed and decided which ones will be applied by using future state BPMN. Then, KPI (Key Performance Indicators) for quality and productivity are evaluated.

In this work an Action-Research (AR) methodology was used in a company of the Manaus Industrial Zone, which is one of the largest manufacturers of HVAC systems and its components in Latin America, and it has capacity to assemble indoor and outdoor units of split systems, as well as the manufacturing of heat exchangers.

AR can be used to improve Statistical Process Control (SPC) firstly recognizing the actual condition, then applying the intended practical action plan, following up the on-going actions, and then monitoring the results, and if necessary, plan a new cycle, starting by the step of recognize actual condition [10].

Manaus industrial park in Brazil has the capacity to assemble HVAC systems as well its components. The process seeks to assure quality of product and process, by monitoring key characteristics, which are measured periodically and controlled by statistical analysis. This process demands resources from the quality department and costs associated need to be controlled and, if possible, reduced.

2 State-of-Art Research

2.1 Refrigeration Systems

ASHRAE [11] defines thermal comfort as "a state of mind that expresses satisfaction with the thermal environment" [11]. Despite being a subjective concept, there are main environmental parameters that influence thermal comfort: temperature, relative humidity, thermal radiation, and average air speed, and other factors such as the clothes worn by people, the level of physical activity performed by them, among other factors [12].

Split HVAC has a part of the system located inside the environment (indoor unit), removing the heat from this environment, filtering, and circulating the air; and another part of the system is in the external area (external unit), being responsible for transferring the heat from the internal environment to the outside.

The heat exchangers are made with copper coils, through the forced passage of air from the environment through the metal fins and they are components responsible for the transfer of heat between fluids at different temperatures, using principles of heat exchange by conduction and convection [13]. And it has specific key characteristics (KC) to be controlled to assure quality of finish good, like Frames per Inch (FPI) quantity and absence of leakage under internal high pressure.

2.2 Industry 4.0 and Manufacturing Data Processing

Industry 4.0 is a concept still under development, which will apply in different ways depending on the reality of a company, but some trends will guide technological developments from now on, being the pillars of Industry 4.0, as big data analysis and monitoring, artificial intelligence, internet of things, robotics, among others [14].

Sarfraz [15] makes a differentiation between concepts of image processing (which is related to image acquisition and its digitalization, for treatment through algorithms), and computer vision (which also studies the theoretical models of the human role for application in artificial systems of data acquisition, processing, analysis and understanding of images to obtain numerical or symbolic information).

Digital image processing systems must have some fundamental steps, where the image is captured (acquisition), imperfections are removed (pre-processing), it is divided into sectors for analysis of a part of interest (segmentation), the sector of interest is analysed through algorithms, filters, artificial intelligence approaches or another method to obtain information (attribute extraction) and finally, the identification of patterns and the storage of the collected information [16].

Another important approach for analysing manufacturing data in the I4.0 is based on simulation, and in [17] is presented a mathematical model through numerical simulation for processing data about fixtures for fork-type parts manufacturing.

There are other important issues regarding manufacturing data processing in the I4.0, among which collaborative approaches, technologies and tools, namely based on the cloud, which are currently of upmost importance [18, 19].

3 Heat Exchangers and SPC Process Improvements

3.1 Process Map and Improvement Opportunities

Heat exchanger manufacturing is composed by main subprocesses below:

3.1.1 U-Bender Tube

This process receives rolls of raw material (copper tubes) and processes to specific dimensions according to product specification. Peg-Leg is defined as length difference in a "U" Shape copper tube, and it is the Critical Characteristics (KC) for this process.

3.1.2 Fin Press

Raw material (steel sheet coils) are cutted in blades and conformed with specific dimensions and shapes. Monitored KCs are the height and diameter of the holes for the tubes.

3.1.3 Package Assembly

Tubes and fins are assembled together in its specified quantities according to heat exchange model. At this point, tubes and fins are only positioned, with no permanent connection between them. There is no data collection at this stage.

3.1.4 Expander

Tube diameters is increased to give the fins and tubes a permanent set through interference fit to allow transfer of heat in final product. The controlled KCs are FPI (number of fins per inch) and tube height.

3.1.5 Brazing

The curves are mounted on the opposite side of the tubes, and then the heat exchanger goes through an automatic welding machine to connect the curves with the tubes through a brazing process. The KC controlled is the absence of fin leakage.

Business Process Model Notation (BPMN) workflow was built to map the process of data collection, as it can be seen in Fig. 1.

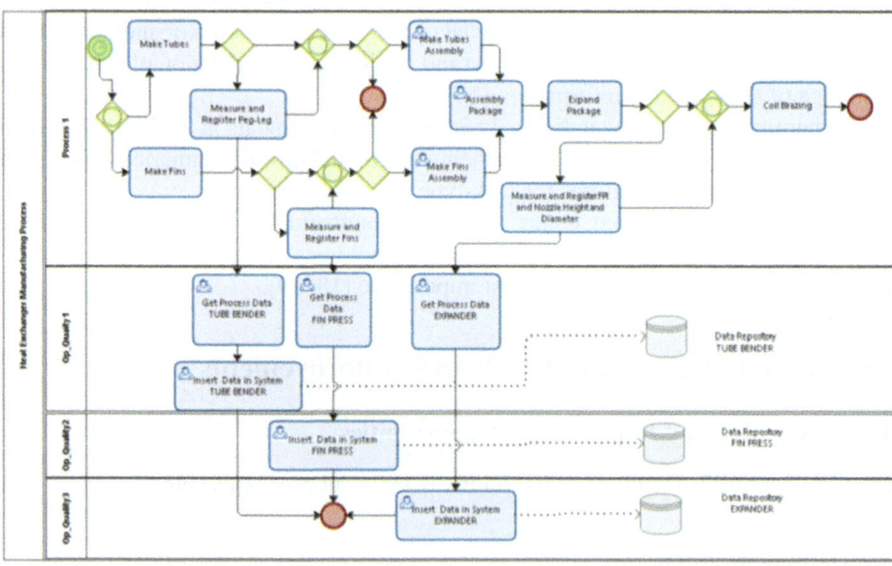

Fig. 1. BPMN of Heat Exchanger production and SPC process, considering actual state.

By analysis of BPMN As-Is, and *in-loco* check, it was possible to identify improvement opportunities, as listed below:

- Some KCs data are not reliable, are not always recorded or are recorded incorrectly.
- Manufacturing operators writes manually data on paper form.
- Quality operator needs to go to process in each machine to collect data. It takes too much time to write all read data on a sheet.

3.2 Scenario 1

Considering the amount of time required for the quality operator to go to the machines and to input data in a repository, the proposed solution to reduce both times was changing the manual control chart to an electronic form where the process operator would input information of measurement on this form.

For the application of scenario 1, electronic forms were developed. The advantages of this solution are the reduction in the processing time of the data measured registration process, the data collection and insertion process in the repository, as well as the paper consumption reduction (paperless concept). These forms were developed using the Microsoft Forms platform, as the example shown in Fig. 2.

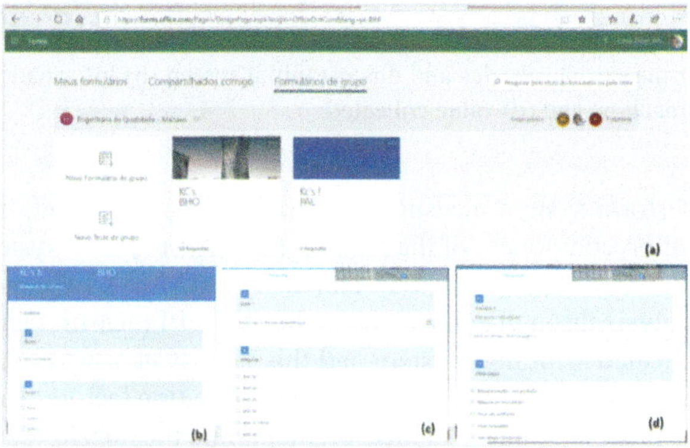

Fig. 2. Electronic form for the collection of cloakrooms and fins (a) Initial screen (b) Information of operator and shift, (c) date and machine and (d) value collected.

3.3 Scenario 2

Due to the expander process characteristics, it was not possible to apply the electronic form immediately. In addition, this process has the biggest processing time and having the most critical characteristics for product performance. An incorrect information will give us wrong conclusions and non-effective actions.

To improve these points, it was suggested to improve the measurement of KC, with the measurement of FPI and nozzles height and diameter through a system of image acquisition and processing with data storage.

Therefore, the system was designed to make the image acquisition by making a comparison with a known standard. In that way, main information is acquired from products. In the case of FPI measurement, height and nozzle diameter, the software has edge pattern detection requirements to accurately obtain the component dimensions, as shown in Fig. 3.

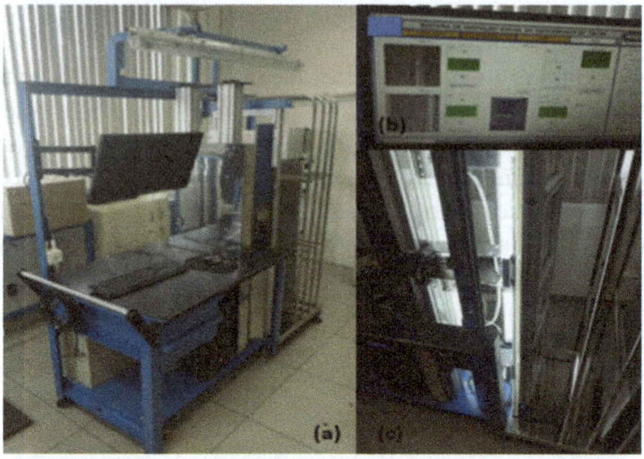

Fig. 3. System for measuring nozzles and fins (a) Initial screen (b) Information of operator and shift, (c) date and machine and (d) value collected.

In order to perform system measurement errors to be evaluated, it is necessary to quantify the variation component of the process and measurement system. So, it is necessary to proceed with a measurement system analysis (MSA), with measurement data collection and analysis through control charts. For the FPI measurement, measurement samples were collected with new system and this new system was compared with the instruments already used today, considering process variation factors. The sampling trees were made for the planning of the FPI measurements and the nozzles diameter, as shown in Fig. 4.

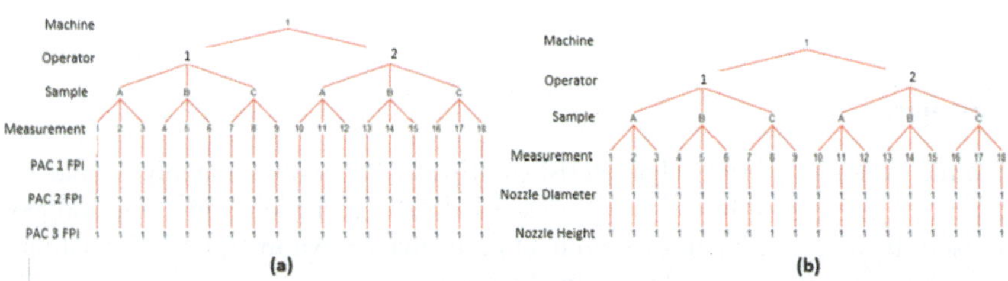

Fig. 4. Sampling trees for data collection (a) FPI (b) Height and diameter of nozzles.

Three samples of heat exchangers were selected, which were measured by 2 operators, each operator taking 3 measurements on each of the instruments. These 3 measures will be used to form subgroups, and in each of these, the mean and variation between the elements are calculated, with the objective of constructing the X-bar R charts and calculating the lower control limit (LCL) and upper control limit (UCL), as shown in Fig. 5.

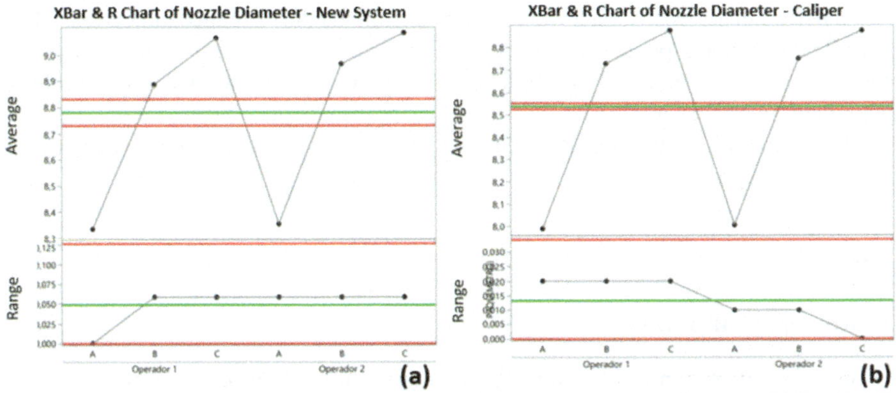

Nozzle Diameter - New System Limit Summaries

Points plotted	LCL	Avg	UCL	Limits Sigma	Subgroup Size
Average	8,732131	8,782444	8,832758	Range	3
Range	0	0,049167	0,126584	Range	3

Nozzle Diameter - Caliper Limit Summaries

Points plotted	LCL	Avg	UCL	Limits Sigma	Subgroup Size
Average	8,526356	8,54	8,553644	Range	3
Range	0	0,013333	0,034328	Range	3

Fig. 5. Comparing before and after improvements (a) diameter of nozzle measured with new system (b) diameter of nozzle measured with caliper.

Range control limits is directly related to measurement error, it was possible to identify that the image inspection system presented a variation in the control limits greater than that measured with the current instrument. Besides this disadvantage, measurement error can be calculated and it is well-known, as shown in Fig. 6.

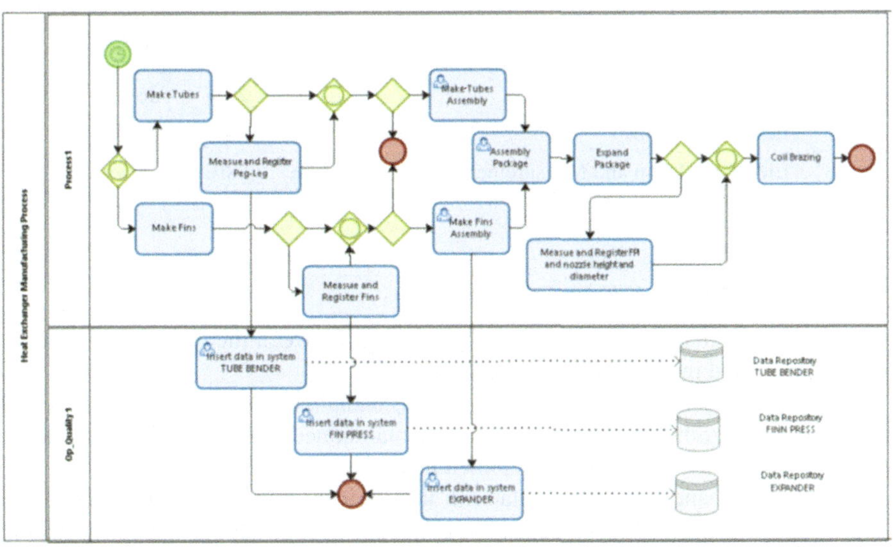

Fig. 6. BPMN of Heat Exchanger production and SPC process, considering future state and applied Scenarios 1 and 2.

In a period of 12 weeks, data were collected from the performance indicators, where weeks 27 to 31 corresponded to the process with no improvement scenarios applied, weeks 32 to 35 correspond to the application of scenario 1 and weeks 36 to 40 refer to the application of scenario 2.

Information about maintenance of related systems [13, 14] was not considered, in detail, in this paper.

4 Results

4.1 Resource Utilization

The resource utilization ratio of quality operators can be defined as the ratio between the sum of all activity times done by operator and the total available time available.

According to Table 1, each scenario application has its contribution for resource utilization decrease, and operators' necessity can be adjusted according to improvements applied in the process. A summary of each simulated scenario and resource utilization can be shown in Table 1.

In Fig. 7, it is possible to see the variation between maximum and minimum observations of occupancy rates in each week and compare with simulation result. The actual values observed from the graphs above were different from the simulated values. This can be explained due to some factors that could not be reproduced, such as variations in the standard time of the activities of operators, limitations of the simulation software that did not allow the group transfer modelling of process components, among others.

Table 1. Summary of simulated results for each scenario.

Legend	Description	Op 1	Op 2	Op 3
C0	No improvements	28%	36%	12%
C1	Electronic Forms	44%	32%	–
C2	Electronic Form and Measurement system by image	38%	–	–

4.2 Leakage Rate

Leak rate in each machine is calculated by the ratio between the quantity of products with leak issue and the quantity produced, in ppm.

After the application of scenario 1, there was a change in the leakage level of both products. However, this level change was only maintained without significant changes after the application of scenario 2, as shown in Table 2.

With the agile analysis of the data with the company main departments involved, it was possible to act in the process at the moment the anomaly occurred, the distinction when the failures are related to machine adjustment or measurement method or tool maintenance, the standardization of operations and the scheduling of preventive maintenance.

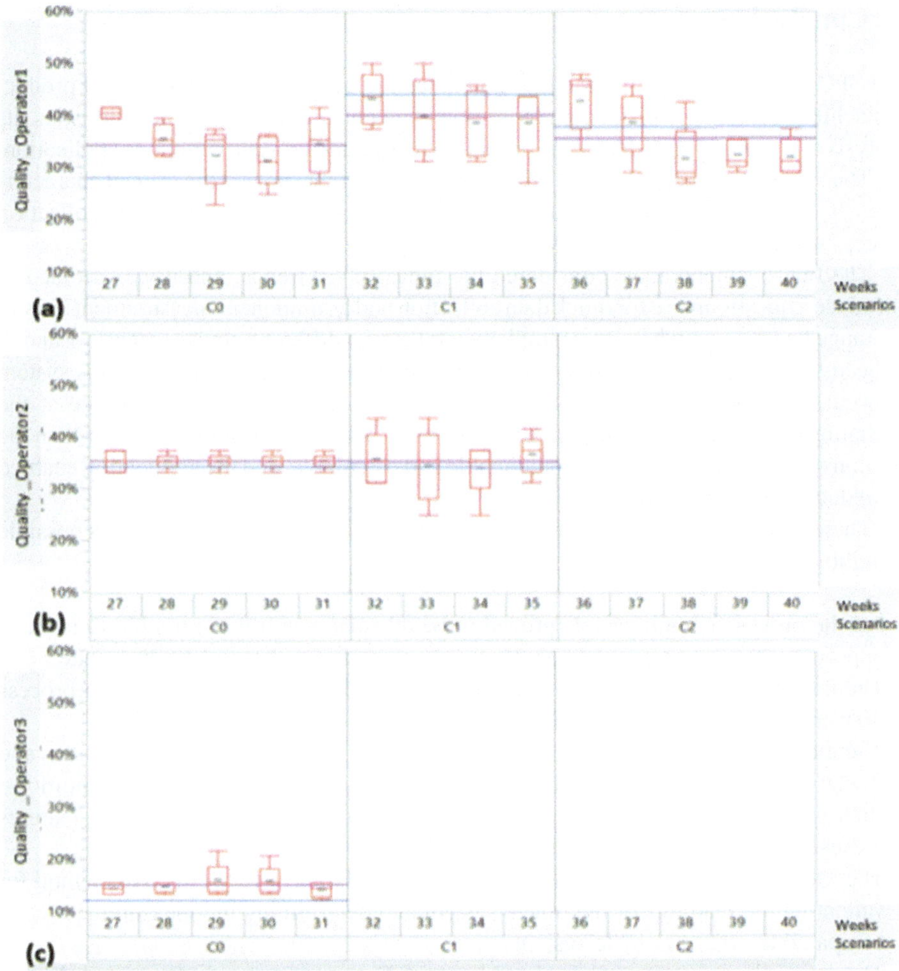

Fig. 7. Observed results of Resource Utilization for (a) Quality Operator 1 (b) Quality Operator 2 and (c) Quality Operator 3.

One of the factors that explained the reduction in the average leakage failure rate in product 2 was the adjustment of the machine and reduction of differences in length between tubes in an agile operation after trends were identified quickly.

Table 2. Leakage Rate with scenarios applied in products 1 and 2 (ppm).

Description	C0	C1	C2
Product 01	2233	962	852
Product 02	3574	1262	1430

5 Conclusions

The SPC process is very important to guarantee the quality of the process and product and achieve customer satisfaction, by sample data collection critical characteristics, its analysis of trends, failure identification and corrective and preventive actions application.

The actual state map was done with a process map that showed the critical characteristics that affects the product's quality and performance, and the indirect data collection process was mapped with the BPMN workflow.

The improvements have considered the Industry 4.0 trends, and it was decided to introduce a questionnaire for quick data collection and a component measurement system by image processing, which were implemented at two different times (scenario 1 and 2, respectively). The data collection systems proved to be very well adapted. This system showed some additional issues such as the excess of corrective maintenance at the beginning of the process operation. Even that error between measures is greater than the previous manual system, they are predictable and calculable, and the application enabled the reduction of process time.

Therefore, as the main points observed in the development of this work, it is possible to highlight:

- All the process mapping tools proved to be adequate, as it was possible to identify opportunities for improvement and to propose an action plan and a future state.
- The use of Bizagi software (which already uses BPMN notation) helped in process flow and suitability for BPM CBOK validation.
- The model was not able to reproduce all operating characteristics of the process and Bizagi software showed some limitations in accurately simulating some conditions, such as the processing of multiple products and the transfer of parts in batches, features present in other available software.
- The Measurement systems Analysis (MSA) proved to be a very effective method for validating the readings of the image system and determining the measurement error compared to other available systems
- The data collection systems based on Microsoft Forms were very well adapted, despite their original objective not to cover this type of application.
- BPMN proved to be adequate as a tool of process map. By its use, it was possible to identify opportunities for improvement and to propose an action plan and a future state.
- The image measurement system was a good resource to improve the SPC process, although there are opportunities for future studies, such as the evaluation of new measurement methods and software improvement for better integration between direct and indirect processes.

Acknowledgement. This work has been supported by national funds through FCT – Fundação para a Ciência e Tecnologia within the project references: UIDB/00319/2020, EXPL/EME-SIS/1224/2021, and UID/CEC/00319/2019.

References

1. Sarti, F., Hiratuka, C.: Desempenho recente da indústria brasileira no contexto de mudanças estruturais domésticas e globais, Para além da política econômica, pp. 127–170 (2018)
2. Bogoviz, A.V., Osipov, V.S., Chistyakova, M.K., Borisov, M.Y.: Comparative analysis of formation of industry 4.0 in developed and developing countries. In: Popkova, E.G., Ragulina, Y.V., Bogoviz, A.V. (eds.) Industry 4.0: Industrial Revolution of the 21st Century. SSDC, vol. 169, pp. 155–164. Springer, Cham (2019). https://doi.org/10.1007/978-3-319-94310-7_15
3. Sakurai, R., Zuchi, J.D.: As revoluções industriais até a Industria 4.0. Rev. Interface Tecnológica **15**(2), 480–491 (2018)
4. Fernandes, N.: Economic effects of coronavirus outbreak (COVID-19) on the world economy. Nuno Fernandes Full Professor of Finance IESE Business School Spain, SSRN Electron. Journal, ISSN 1556-5068, Elsevier BV, pp. 0–29 (2020)
5. Lizarelli, F.L., Bessi, N.C., Oprime, P.C., Amaral, R.M., Chakraborti, S.: A bibliometric analysis of 50 years of worldwide research on statistical process control. Gestão Produção **23**(4), 853–870 (2016)
6. Petrovic, S., Milosavljevic, Sajic, L.: Rapid evaluation of maintenance process using statistical process control and simulation Int. J. Simul. Model. **17**, 119–132 (2018)
7. Gibbs, L., Shea, T.: The Power of Two The Power of Two: Combining Lean Six Sigma and BPM (2007)
8. Antosz, K., Chandima Ratnayake, R.M.: Machinery classification and prioritization: empirical models and AHP based approach for effective preventive maintenance. In: IEEE International Conference on Industrial Engineering and Engineering Management, art. no. 7798104, pp. 1380–1386 (2016). https://doi.org/10.1109/IEEM.2016.7798104
9. Jasiulewicz–Kaczmarek, M., Antosz, K., Żywica, P., Mazurkiewicz, D., Sun, B., Ren, Y.: Framework of machine criticality assessment with criteria interactions. Eksploatacja i Niezawodnosc **23**(2), 207–220 (2016). https://doi.org/10.17531/ein.2021.2.1
10. Coughlan, P., Coghlan, D.: Action research for operations management. Int. J. Oper. Prod. Manag. **22**(2), 220–240 (2002). https://doi.org/10.1108/01443570210417515
11. ASHRAE Thermal Environmental Conditions for Human Occupancy, ASHRAE Stand., no. 55 (2003)
12. Abrava, Recomendação Normativa ABRAVA RN 03 - 2003 - Sistemas de Condicionamento de ar para conforto Parâmetros de Conforto Termico (2003)
13. Moran, M., Shapiro, H., Munson, B., DeWitt, D.: Introdução à Engenharia de Sistemas Termicos: termodinâmica, mecânica dos fluidos e transferencia de calor, 1st ed. Grupo Gen-LTC, Rio de Janeiro (2011)
14. World Economic Forum: Deep shift: technology tipping points and societal impact World Econ. Forum, no. September, pp. 1–44 (2015)
15. Sarfraz, M.: Intelligent Computer Vision and Image Processing: Innovation, Application, and Design. IGI Global (2013)
16. Gonzalez, R., Woods, R., Eddins, S.: Digital Image Processing, 3rd edn. Gatesmark Publishing, Knoxville (2009)
17. Ivanov, V., Pavlenko, I., Kuric, I., Kosov, M.: Mathematical modeling and numerical simulation of fixtures for fork-type parts manufacturing. In: Knapčíková, L., Balog, M. (eds.) Industry 4.0: Trends in Management of Intelligent Manufacturing Systems. EICC, pp. 133–142. Springer, Cham (2019). https://doi.org/10.1007/978-3-030-14011-3_12

18. Varela, M.L., Putnik, G.D., Manupati, V.K., Rajyalakshmi, G., Trojanowska, J., Machado, J.: Collaborative manufacturing based on cloud, and on other I4. 0 oriented principles and technologies: a systematic literature review and reflections. Manag. Prod. Eng. Rev. **9** (2018). https://doi.org/10.24425/119538
19. Ferreira, L., et al.: Framework for collaborative practices platforms for humans and machines in industry 4.0 oriented smart and sustainable manufacturing environments. In: Manupati, V.K., Goran, D.P., Rocha, M.L. (eds.) Smart and Sustainable Manufacturing Systems for Industry 4.0. CRC Press, Taylor & Francis Group, Boca Raton (2022, in Press)

Analysing Impact of the Digitalization on Visual Inspection Process in Smartphone Manufacturing by Using Computer Vision

Josilene Lima[1](✉), Vaibhav Shah[2,4], Leonilde Varela[1,4], Caetano Monteiro[3,4], Goran Putnik[1,4], and José Machado[3,5]

[1] Department of Production and Systems, Universidade do Minho, Guimarães, Portugal
josyfelima@gmail.com
[2] Department of Information Systems, Universidade do Minho, Guimarães, Portugal
[3] Department of Mechanical Engineering, Universidade do Minho, Guimarães, Portugal
[4] Centro ALGORITMI, Universidade do Minho, Guimarães, Portugal
[5] MEtRICs Research Center, Universidade do Minho, Guimarães, Portugal

Abstract. This work shows a case study of the application of data digitalization in visual inspections on smartphone manufacturing, from the context of the need for inspection carried out by humans, in order to analyze its impacts. The impacts of visual stress and fatigue in employees, the technological techniques that can be used through computer vision and artificial intelligence to reduce those impacts in a person, along with the support that digitization can bring to the daily lives of companies and workers are focused in this paper. In this paper is also shown how the digitalization actions were proposed and applied, using a new method for prioritization, along with preliminary results obtained.

Keywords: Superficial inspections · Digitalization · Computer vision · Artificial intelligence

1 Introduction

The use of smartphones in the world has become common in people's lives. These contain several features, from the simplest to the most complex, such as making calls and typing documents resembling notebooks. It is considered that "Currently there is a great interest of an organization in guaranteeing the quality of its products and services in order to remain in the consumer market" [1].

In Brazil, to carry out an analysis of the quality of products, the standards defined in technical standard 5426 must be followed to obtain, test, and/or compare to verify the list of possible defects according to their severity and identification of what is called AQL (Acceptable Quality Level) [2].

The AQL parameters also assess the superficial condition of smartphones during manufacturing. These are performed visually by employee in the productive environment, through a visual inspection process [3, 4] but this visual inspection can impact the worker's visual stress because of a long periodicity of this activity [5].

A. Hamrol et al. (Eds.): MANUFACTURING 2022, LNME, pp. 125–137, 2022.
https://doi.org/10.1007/978-3-031-00218-2_11

Looking at the study done by Mora [6], it shows a sample of the main symptoms related to visual discomfort. This approach was defined to distinguish the different symptoms associated with this problem, namely the visual fatigue, blurred vision, visual irritability, headaches, stress, and difficulty concentrating. It is interesting to note that this portrays activities in the office, but it can lead to a parallel reference on what it would be like to work with vision in a factory environment.

For Mora [6], his study showed that for people who work in offices that 50% of employees wear glasses and 60% have ophthalmological problems, such as myopia and astigmatism, thus generating greater visual fatigue. The prevalence of the visual fatigue is expressed in percentage, relatively to the total number of the participants. In this case, 30% of respondents feel visual fatigue 17% have visual fatigue at the beginning of the day, 33% in the middle of the day and 50% at the end of the day. These results indicate that after several hours of continued work with few pauses, it can lead to long-term visual problems.

Among the types of symptoms that can be generated with the use of prolonged human vision, there are terms called fatigue and visual stress and for Pimenta et al. [7], fatigue is as one of the main causes of human error. Many times, its symptoms are ignored, as well as its importance for a good mental and physical condition, elementary for human performance and health. Fatigue is however a very subjective concept and difficult to define from a scientific point of view. It may be a combination of symptoms that include loss in performance.

During technological advances, methods of intelligent computing systems based on images like human vision, capable of being processed through computers, have emerged, according to Szeliski, 2010 (as cited Araújo [8]). These become alternatives for dealing with impacts related to visual stress in long periods and fatigue.

The technological fronts that can contribute to reduce impacts of visual stress and fatigue are computer vision & Artificial Intelligence (AI) through computer vision software and AI algorithms in machines. It has been defined as the science and engineering of making intelligent machines and as human-like intelligence, either in machines or software. AI technologies are used widely today to optimize processes, to make products easier to use, and to automate tasks. Within AI research, we look for new methods of solving problems, based on action-research approach [9]. This search generates challenges in terms of knowledge representation and reasoning Introduction 35 methods, planning, learning, natural language processing, motion, manipulation, perception, social and evolutionary intelligence, feelings, and creativity. These techniques can be applied and used in areas such as medicine, psychology, weather, finance, transportation, gaming, aviation, and in the law, where they have been applied. Many of these advances have been targeted at vehicle drivers [10].

This paper is organizing the following demonstrating the context of the importance of Smartphone manufacturing in society associated with good quality of human visual inspection and impacts, some technologies digitalization that can reduce these impacts, the application of the computer vision in the manufacture and conclusions.

2 Context

Today there is a use of approximately 355 million smartphones produced in the first quarter in the world, data from 2021, and massively manufactured by around 7 large companies, including: Samsung, Huawei, Apple, Xiaomi, Oppo, Vivo, Motorola. These data reflect the reality of a demanding market [11] in Fig. 1. This paper was analysed in a company from the Industrial Pole of Manaus, Brasil, in smartphone manufacturing.

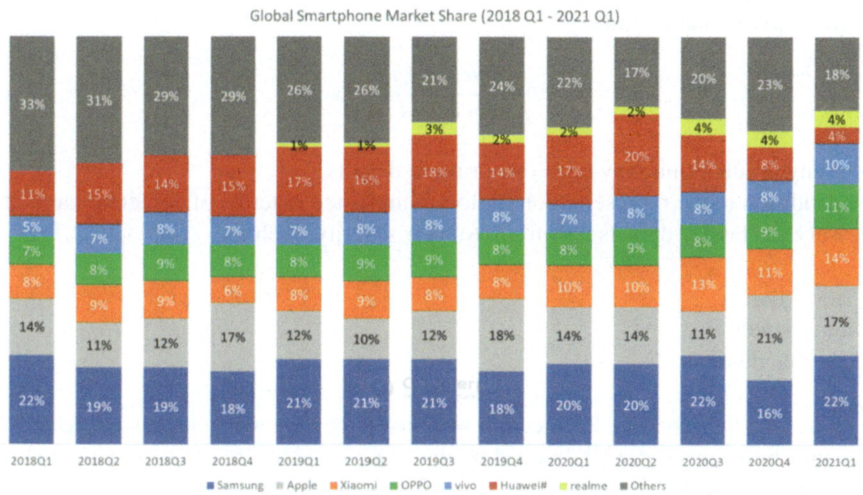

Fig. 1. Smartphone production in the world.

For Lobato [1], the market requires companies to produce quality products at a lower cost. To achieve results in this direction, it is necessary to maintain a dedication to constant improvements.

Looking at the course of history, for Pinheiro et al. [12], "quality went through four stages of development: inspection, process supervision, control and strategic quality management".

"The first, the quality inspection phase, in which the final products were examined based on visual inspection, separating the products with defects that should be destroyed or return to the production process for correction" [13].

To understand a little more, about how superficial defect analysis has been done, the next section presents a brief description of the product quality analysis carried out in this work.

2.1 Product Quality Analysis Performed by People in the Production Environment

To analyse the quality of products, standards established in technical standard 5426 must be followed to obtain measurements, tests, and/or comparisons to verify the list of possible defects according to their severity and identification of what is called AQL [2].

Within the AQL parameters it is necessary to find the non-Conformity in the day-to-day and this is expressed in terms of "percent defective" or in terms of "defects per hundred units".

- Defective percentage:

$$= \frac{Defective\ Units}{Inspectiined\ unit} x\ 100 \tag{1}$$

- Defects per hundred Units

$$= \frac{Defective\ Units}{Inspectiined\ unit} x\ 100 \tag{2}$$

** any product unit may have one or more defects.

In addition to the representation of non-compliance in terms of defective parts, it is necessary to classify defects according to their severity, such as:

- Critical defect;
- Major defect;
- Tolerable defect.

Therefore, the superficial inspection work itself, being a work entirely carried out by the human being, even if well directed by the AQL, is subject to errors at the time of its application. Thinking about it, and looking at existing technologies, there are currently computer vision techniques associated with Artificial Intelligence that can minimize these impacts, as can realised through the information summarized below.

2.2 Digitalization with Computer Vision and Artificial Intelligence Techniques

According to the Data Science Academy team, published in 2018 [14], computer vision is the process of modelling and replicating human vision using software and hardware that studies how to reconstruct, interrupt, and understand a 3d scene from its 2d images in terms of its properties of the structure present in the scene.

Therefore, the superficial inspection work itself, which guarantees quality, being carried out by the human being, and even if well directed by the AQL, is subject to errors at the time of its application. Considering this, and looking at existing technologies, there are currently techniques of computer vision system (SVC). With the technology advancing, methods of intelligent computer systems based on images, like human vision have emerged, capable of being processed through computers, or computer vision. In the context of the presented case study, i.e. detection of visual superficial defects on smartphone devices, there are multiple works that demonstrate various machine learning based approaches to detect smartphone surface defects, through computer vision [15]. The most common superficial defects are shown in Fig. 2 (Fig. 3).

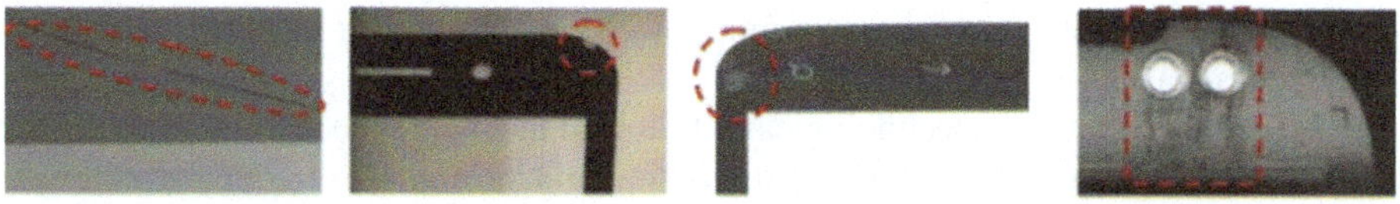

(a) Scratch (b) Chip (c) Dirt (d) Light leak

Fig. 2. Common superficial defects as 'areas of interest'.

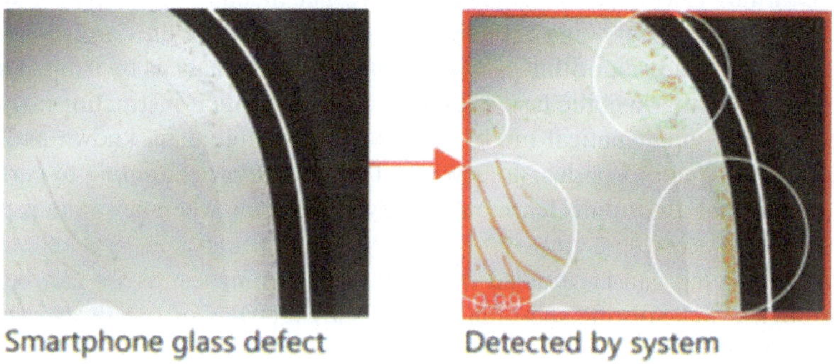

Fig. 3. Automatic detection of 'areas of interest' with smartphone surface defects.

Computer vision can be used to digitize social media platforms to find relevant images that cannot be discovered through traditional searches. The technology is complex and, like all the tasks mentioned above, it requires more than just image recognition, but also semantic analysis of large data sets. Data Science published in 2018 [14].

Computer vision systems (SVC) started in the 70s with approaches in Artificial Intelligence (AI) according to Lopes [16]. It is a technology that has been gaining strength and giving directions to image processing, using cameras, optical sensors, scanners, among others [16].

It can be said that AI aims to perform and simulate human functions such as the ability to learn, adapt and make decisions in different situations according to knowledge according Oliveira [17].

In the field of AI, one can cite the branch named machine learning (ML), or also known as machine learning, which intends to perform human functions for learning with provisioning and decision-making [18].

Machine learning is a subset of AI applications capable of learning on their own. In fact, it reprograms itself, as it digests more data, to perform the specific task for which it was designed with increasing precision [19].

In this article ML is used to provide surface analysis on smartphones through computer vision and Artificial Intelligence techniques, but there are other 6 powerful, and frequently mentioned use cases for machine learning in manufacturing, which are: 1- predictive maintenance, 2 - predictive quality and throughput, 3 - digital twins, 4 - generative design/intelligent manufacturing, 5 - energy consumption prediction and 6

- cognitive supply chain management (six powerful use cases for machine learning in manufacturing) [20–22].

These approaches have technologically followed the main objectives underlying the Industry 4.0 (I4.0), which is aimed at digital manufacturing, also frequently called as smart factory, further associated to intelligent networks, mobility, flexibility of industrial operations and its interoperability" and collaboration [23–25].

In a general approach context, Artificial Intelligence is the area of computer science that emphasizes the creation of intelligent machines that work and react like humans. Machine Learning is a branch of Artificial Intelligence that allows computers to learn to perform new tasks without being explicitly pre-programmed for the corresponding end. Thus, it enables to study algorithms and perform statistical analysis by using computers to accurately perform a specific task without the use of explicit instructions.

In this work a mathematical model was built on sample data, known as "training data", to make predictions or decisions without any direct programming to perform the task. These types of algorithms learn on their own and grow when new data is provided for the machine to learn.

Machine learning can be divided into three main branches, as we can be seen in Fig. 4, through which specific sub-branches are shown [10].

The sub-branches are:

- Supervised Learning.
- Unsupervised Learning.
- Reinforcement Learning.

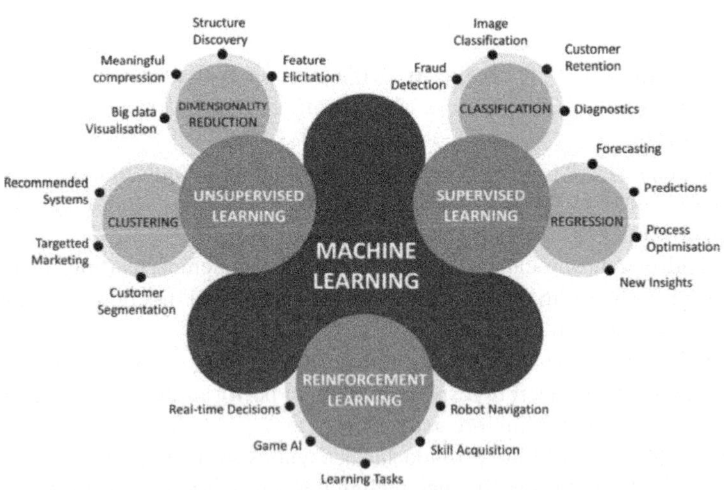

Fig. 4. Machine learning in its three main sub-branches [10].

2.2.1 Supervised Learning

In supervised learning, the objective is stated with labelled/classified training data. There is a value, an assigned class, that allows you to have an idea of what you are looking to learn. That training data consists of the input vector X and output vector Y with labels. A vector Y label is the explanation of its respective input vector X data. Together, they form a training example that can describe a relationship that makes sense.

For example, as a child, parents teach their children about the names (labels) of objects by pointing at them and pronouncing their names, in a supervised manner.

2.2.2 Unsupervised Learning

In unsupervised learning, there is no supervisor indicating/labelling a vector with a defined objective and therefore there is no data for training. There is no clarity about what you want to learn. This is a fact that data scientists and machine learning professionals often find, where a large amount of data is available, but without labels. Still, it is possible to learn from the search for significant or hidden structures or behaviours in the data [10].

2.2.3 Reinforcement Learning

According to Deep Learning Book [26] the Reinforcement Learning, is the training of machine learning models to make a sequence of decisions. The agent learns to achieve a goal in an uncertain and potentially complex environment. In reinforcement learning, the artificial intelligence system faces a situation. The computer uses trial and error to find a solution to the problem. For the machine to do what the programmer wants, artificial intelligence receives rewards or penalties for the actions it performs. Your goal is to maximize the total reward. We can see below the Fig. 5.

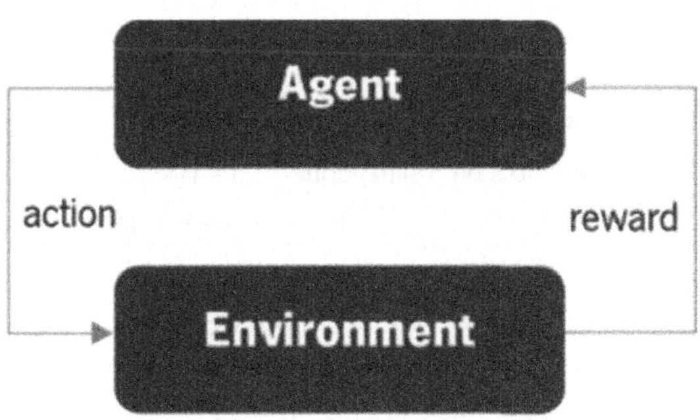

Fig. 5. Reinforcement learning process.

All branches lead to machine learning and show approaches that can be applied to consider what type of data is input and which answer is sought in the output of the modelled algorithm. Thus, it is necessary to analyse what type of data is generated in the smartphones' surface defects inspection through a vision system (system input data) and which metrics are sought in the output.

3 Smartphone Production Quality Control in the Company

The general processes of the smartphone company are related according to the macro layout expressed in Fig. 6, and an explanation of each function, in Table 1. In the image, we can see where the superficial inspection process is, this at post 5, and in addition, there is a sample, at post 7, for review to review in a lower percentage if the smartphones have a good surface.

By analysis of BPMN As-Is, and *in-loco* check, it was possible to identify improvement opportunities, as listed below:

- The data collected from some KCs is not always reliable and are not always recorded or are recorded incorrectly.
- Manufacturing operators manually writes data on paper.
- Quality operator needs to go to each machine to collect data and takes too long to manually write all data collected on a spreadsheet.

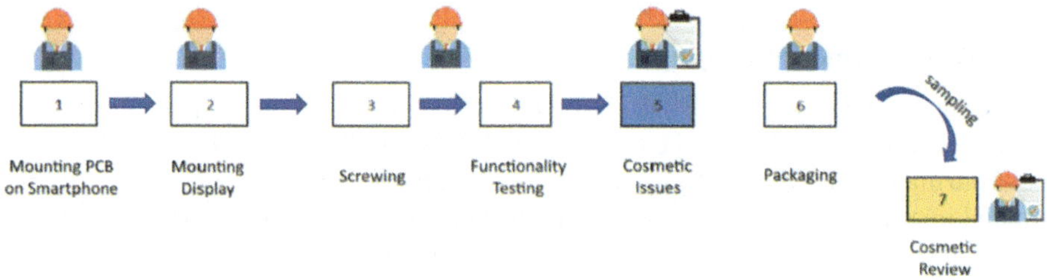

Fig. 6. Macro smartphone assembly process.

Table 1. Macro details of mounting smartphones (own creation)

Process number	Process name	Assy
1	Mounting PCBS on Smartphone	PCBS, battery, cables, Labels
2	Mount display	Screen, labels, microphone, camera
3	Screwing	Closing
4	Function tests	Colour test, screen, audio, network
5	Surface issues	Check Strokes, scratches, impurities
6	Packing	Mount phone and accessories in the box
7	Surface review	Review surface (~5%)

During the monitoring of the process, in the period of 6 months of superficial inspection on smartphones (inspection of smartphones' surface quality, whether there is any visible defect in appearance, such as scratches etc.), carried out by human operators (inspectors), it was observed that there were defective pieces that passes through the operators as if they were "good" products and, similarly good products, i.e. without any visible surface defects were labelled to be defective by the human operators. The portions of good quality product, and superficially defective product, out of all the smartphones manufactured in the company, and sampled at Station 7, are highlighted in Table 1. This index is based on the behavioural average of monthly production of the company. Looking at the graph in Fig. 7, which represents the inspection performed by the human operators, it is evident that there were failures that occurred in the judgment of the inspection station, the station 5. This could have been caused by several factors. Considering the data collected, the biggest challenge is to reduce the **1.67%** of **% false negatives**, i.e., the undetected 'defective' phones. This could mean that the system may pass some superficially defective phones to consumers.

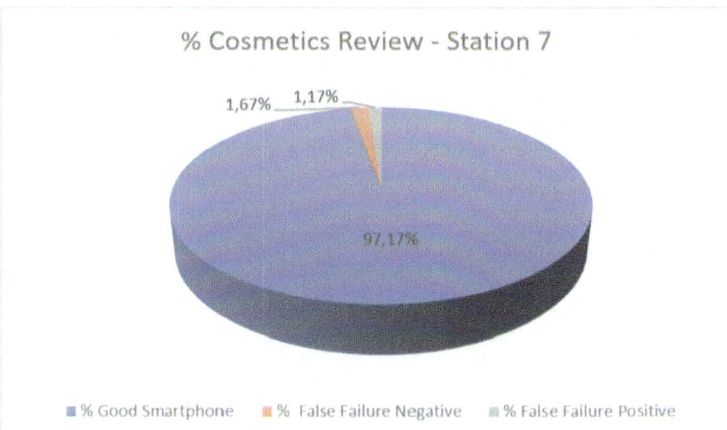

Fig. 7. Percentage of surface inspection – station 7 (own creation).

However, in this company there is another filter that reviews the smartphones again before packaging. Therefore, the main objective of the company was to guarantee that it would not pass superficial defects in production, without the need for a second revision before packaging the products. Hence, it installed an automatic surface inspection machine to recognize superficial defects by the capturing images, through computer vision assisted by Artificial Intelligence using Reinforcement learning and Neural Networks (NN) algorithms. In this case, the model is trained with an image bank of surface defects. However, the company had not yet made an efficiency comparison between both alternatives (with and without the second inspection phase), to be able to clearly conclude whether the result of the completely automated station would be meeting its needs. Therefore, a case study was considered for comparing the efficiencies of both alternatives, the AI station and the human based solution, which will be briefly discussed below.

4 Results Discussion and Analysis

During a period of 20 days of smartphone production in the factory, a total of 1824 samples were collected for the present study. A machine with an automatic surface inspection was used to recognize superficial defects through the captured images, based on computer vision assisted by Artificial Intelligence using Reinforcement learning and Neural Networks (NN) algorithms. There were produced 3 batches, to compare the efficiency of the operator and the AI station. The objective was to understand the effectiveness of using artificial intelligence in the day-to-day processes in the production facility, to help employees in the detection of surface defects in the smartphones. These passed at the AI station in a similar way as those evaluated by the human operators, without one knowing what the other has identified. Again, since the objective of the inspection was to detect a defect, the samples with defect were labelled as 'positives' and the others are 'negatives'.

After the period of collection and evaluation by both, the operator and the AI station, the results have been compiled, as summarized in Table 2.

Table 2. Smartphone's inspection results – confusion matrix data for both inspection methods

	Without defect		With defects		Total
	True negatives	False positives	True positives	False negatives	
Manual inspection	1557	12	194	61	1824
AI station	1371	198	255	0	1824

To evaluate and compare the two inspection methods, the values from the confusion matrix for each method from Table 2 were then applied to calculate the AI system evaluation metrics as in previous examples. The values of these metrics are presented in Table 3.

Table 3. Results of the analysis with the AI parameters with data from the operator and the AI station

	Accuracy	Precision	%False positives	%False negatives	Recall
Manual inspection	96.00%	94.17%	0.66%	3.34%	76.08%
AI station	89.14%	56.29%	10.86%	0.00%	100.00%

Based on the results of Table 3, it is necessary to understand what the client asked for as a priority in Business and data understanding, which was to avoid passing surface defects and this is represented by what we call recall, that is, the greater the recall the greater the capacity to identify defects. Considering what has been measured, the AI station has a greater capacity to detect defects as compared to the human operators' powered station. This helps to achieve the objective of this company. Although the operator had greater accuracy, this does not represent the result that the client expects. This may have occurred due to visual stress of the worker, due to a long periodicity of this activity or the changing of operator in the 3 shifts, or some other issues which are not the scope of this study.

However, the AI powered station helped clearly to achieve the goal, under the given circumstances. In other words, after the comparison of the results in terms of evaluation parameters from both methods, it may be declared helpful to have the AI-powered station to help employees/inspectors in the detection of surface defects in daily routine of the production process. Besides providing better quality and efficiency, the AI powered station may help reducing the cost of quality control as well.

5 Conclusions

This work aimed at presenting the basic concepts in machine learning, showing its historical background and how the ML algorithms can be evaluated and applied in an industrial context, more precisely in smartphones surface quality inspection process. The main purpose was to enable a digital transformation in the factory, along with an improvement on production productivity, and in quality, along with costs reduction.

The case study considered for the application of data digitization analysis on superficial inspections carried out in smartphone manufacturing system, which did enable to realize about the support that digitization can bring to the daily lives of the company and its workers.

The main objective was achieved, through the application of machine learning in a world industrial scenario, through the application of data digitization analysis on smartphones' surface visual inspection process in the considered company, which did enable to improve the process efficiency, and further in terms or production management, and quality control, while reducing costs. Therefore, the use of these kind of technology and approaches enable the manufacturing company to be more efficient and remain innovative and competitive. The main achievements are thus mainly related to, better quality control, higher production volumes, reduction of human stress due to repetitive labour (with occupational risks and avoiding human-induced errors), and gaining more accuracy through artificial intelligence powered machines with human support, thus achieving better general results.

Contrary to the third industrial revolution, which was a digital revolution, where computers started replacing humans for many 'intelligent' tasks, the current fourth revolution considers the 'humans' in the centre, despite some of the 'intelligence-intensive', 'smart' and 'automatic decision-making' tasks that have been being passed to the machines, as emphasized in [21, 22], regarding the two-part issues, related to the I4.0, and to the underlying importance of collaboration [25].

Acknowledgement. This work has been supported by national funds through FCT – Fundação para a Ciência e Tecnologia within the project references: UIDB/00319/2020, EXPL/EME-SIS/1224/2021, and UID/CEC/00319/2019.

References

1. Lobato, T.T.: O Sistema Kaizen Como Alicerce Para o Lean Manufacturing: O Caso de Um Centro de Distribuição de Uma Empresa de Cosméticos, 69 (2019)
2. ABNT NBR 5426. Sampling Plans and Procedures in Attribute Inspection (2015)
3. Kuric, I., Kandera, M., Klarák, J., Ivanov, V., Więcek, D.: Visual product inspection based on deep learning methods. In: Tonkonogyi, V., et al. (eds.) InterPartner 2019. LNME, pp. 148–156. Springer, Cham (2020). https://doi.org/10.1007/978-3-030-40724-7_15
4. Kujawinska, A., Vogt, K., Diering, M., Rogalewicz, M., Waigaonkar, S.D.: Organization of visual inspection and its impact on the effectiveness of inspection. In: Hamrol, A., Ciszak, O., Legutko, S., Jurczyk, M. (eds.) Advances in Manufacturing. LNME, pp. 899–909. Springer, Cham (2018). https://doi.org/10.1007/978-3-319-68619-6_87
5. Vilhena, D., Freitas, S., Guimarães, M., Pinheiro, A.: O papel do psicopedagogo na identificação e intervenção nos distúrbios de aprendizagem relacionados à visão : caso de uma intervenção tardia. O Papel Do Psicopedagogo Na Identificação e Intervenção Nos Distúrbios de Aprendizagem Relacionados à Visão: Caso de Uma Intervenção, 49 (2018)
6. Mora, J.A.: Study of risk factors that influence visual fatigue and musculoskeletal stress in an open office Work done under the academic supervision of Ana Sofia de Pinho Colim (2019)
7. Pimenta, A., Carneiro, D., Novais, P., Neves, J.: Monitoring mental fatigue through the analysis of keyboard and mouse interaction patterns. In: Pan, J.S., Polycarpou, M.M., Woźniak, M., de Carvalho, A.C.P.L.F., Quintián, H., Corchado, E. (eds.) Hybrid Artificial Intelligent Systems, pp. 222–231. Springer, Heidelberg (2013). https://doi.org/10.1007/978-3-642-40846-5_23
8. Araújo, P.D.: Análise e classificação da qualidade do Linter e do óleo de algodão utilizando técnicas de visão computacional (2018)
9. Coughlan, P., Coghlan, D.: Action research for operations management. Int. J. Oper. Prod. Manag. **22**(2), 220–240 (2002). https://doi.org/10.1108/01443570210417515
10. Ribeiro, T.A.O.: Deep Reinforcement Learning for Robot Navigation Systems. Universidade do Minho, Escola de Engenharia, Portugal (2019)
11. CounterPoint: Global Smartphone Market Share: By Quarter (2021). https://www.counterpointresearch.com/global-smartphone-share/. Accessed 15 July 2021
12. Pinheiro, R., Viaro, F., Teixeira, F., Silva, R.: Aplicativo de Desdobramento das Funções da Qualidade (QFD) Utilizando Conceitos de Programação Orientada a Objetos. Aplicativo de Desdobramento Das Funções Da Qualidade (QFD) Utilizando Conceitos de Programa Orientada a Objetos, 15 (2018)
13. Sousa, R.D.O.: Qualidade na Administração Pública: o impacto da certificação ISO 9001: 2000 na satisfação dos munícipes, pp. 1–121 (2007). http://repositorium.sdum.uminho.pt/handle/1822/7020
14. Data Science. publicado O que é visão computacional? O Que é Visão Computacional? - Data Science Academy (2018). http://datascienceacademy.com.br/blog/o-que-e-visao-computacional/
15. Jian, C., Gao, J., Ao, Y.: Automatic surface defect detection for mobile phone screen glass based on machine vision. Appl. Soft Comput. J. **52**, 348–358 (2017). https://doi.org/10.1016/j.asoc.2016.10.030
16. Lopes, F.: Visão computacional para estimativa de comportamento de aglomeração de galinhas poedeiras, 72 (2018)
17. Oliveira, D.: Um sistema inteligente que prevê a produtividades do algodão em imagens de lavouras comerciais, 56 (2019)
18. Baptista, D.: Machine learning approaches for predicting effects of drug combinations in cancer. (June), 77 (2016)

19. Six Powerful Use Cases for Machine Learning in Manufacturing (eleks.com), 5th May 2021. https://eleks.com/blog/machine-learning-in-manufacturing/
20. Shah, V., Costa, D.E.B., Moreira, S.F., Lima, J.F., Varela, M.L.R., Putnik, G.D.: Machine learning applications for industry 4.0. In: Manupati, V.K., Putnik, G.D., Varela, M.L.R. (eds.) Smart and Sustainable Manufacturing Systems for Industry 4.0. CRC Press, Taylor & Francis Group (in press)
21. Putnik, G.D., Shah, V., Putnik, Z., Ferreira, L.: Machine learning in cyber-physical systems and manufacturing singularity – it does not mean total automation, human is still in the centre: part II – in-CPS and a view from community on industry 4.0 impact on society. J. Mach. Eng. **21**(1), 133–153 (2021a). https://doi.org/10.36897/jme/134245
22. Putnik, G.D., Pabba, S.K., Manupati, V.K., Varela, M.L.R., Ferreira, F.: Semi-double-loop machine learning based CPS approach for predictive maintenance in manufacturing system based on machine status indications. CIRP Ann. Manuf. Technol. **70**(1), 365–368 (2021). ISSN 0007-8506. https://doi.org/10.1016/j.cirp.2021.04.046
23. Barreto, L., Amaral, A., Pereira, T.: Industry 4.0 implications in logistics: an overview. Procedia Manuf. **13**, 1245–1252 (2017)
24. Xu, L.D., Xu, E.L., Li, L.: Industry 4.0: state of the art and future trends. Int. J. Prod. Res. **56**, 2941–2962 (2018)
25. Ferreira, L., et al.: A framework for collaborative practices platforms for humans and machines in industry 4.0 oriented smart and sustainable manufacturing environments. In: Manupati, V.K., Goran, D.P., Rocha, M.L. (eds.) Smart and Sustainable Manufacturing Systems for Industry 4.0. CRC Press, Taylor & Francis Group, Boca Raton (2022, in press)
26. Deep Learning Book. Visto 06/03/2021. O que é visão computacional? Capítulo 62 - O Que é Aprendizagem Por Reforço? - Deep Learning Book. http://deeplearningbook.com.br/o-que-e-aprendizagem-por-reforco/

Design of a Vision System for Needles' Beds Positioning Inspection: An Industrial Application

Filipe Pereira, Luis Freitas, Rui Oliveira, José Vicente, Teresa Malheiro, Arminda Manuela Gonçalves, and José Machado(✉)

University of Minho Campus of Azurém, 4800-058 Guimaraes, Portugal
d8561@dem.uminho.pt

Abstract. Printed Circuit Boards (PCBs) are components with increasing use for many applications. The industrial production of these components led to a situation of necessary efficient automation performance, for their production, and, mainly, of test systems for their quality control during the production process. Quality control of PCBs is a very complex task and the development of automatic testing machines is an asset for this purpose. This work is focused on the development of a vision system to be used in an automatic machine for testing PCBs in a systematic and automated procedure. The test procedure developed for PCBs is based on the development of a tool composed of needles that, when coming into contact with PCBs at predetermined contact points, allow the measurement of different predefined parameters and conclusion on their quality. The task of checking and controlling the quality of these needles, in the developed tool, is also automated, requiring a vision system to ensure that these needles meet the requirements for their assembly. This vision system is capable of detecting problems in the needle beds, and it is important to assess whether they will be able to be embedded in the PCBs, taking this process into account.

Keywords: Vision system · Robotic system · Mechatronic design · Needles' beds · Positioning systems · Inspection systems

1 Introduction

Technological production is growing today at a very high rate, which drives electronic equipment manufacturers into a race for the supply of electronic components at a high rate. Printed circuit boards (PCBs) are an electronic component widely used globally and represent a highly developed market [1].

Increasingly, PCB boards have a reduced size and an increase in electronic components on them one of the main problems affecting the production of PCBs is the fact that there are not many quality analysis systems in the electrical quality testing phase (circuit test systems - ICT systems) in the automotive industry [1].

This lack of analysis/quality check leads to plates being rejected or left on standby in order to be repaired at a later stage [1].

A. Hamrol et al. (Eds.): MANUFACTURING 2022, LNME, pp. 138–153, 2022.
https://doi.org/10.1007/978-3-031-00218-2_12

The big problem lies in the fact that PCBs are still mostly produced by hand. This leads to failures in the manufacture of ICT systems in the final product. Often these needles can suffer damage caused by the handling systems or transport systems of the same [1].

ICT systems consist of a test plate (DUT) and a set of needles that are placed on it. The needle bed consists of a test plate that can be of different sizes and has several holes. Here in these holes the receptacles and needles will be inserted. On the other side of the test plate, the set consisting of the receptacle and the needle, is coupled to a wiring system for electrical connections.

This wiring system allows electrical parameters to be measured on the PCB and a connection between software and printed circuit board [1]. Figure 1 shows each side of the test plate in detail: the needle bed and the wiring system, respectively [1].

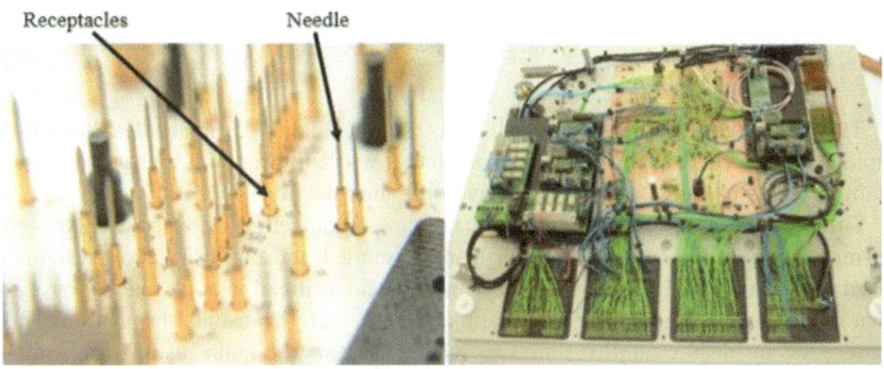

Fig. 1. Test plate [1].

The needles are manually inserted into the receptacles, which serve as a guide for the needles (see Fig. 2). This manual manufacturing process can take longer than a week depending on the type of PCB being tested. This being a repetitive process that requires operators to have a high sense of responsibility and level of concentration, mistakes can happen with the wrong choice of needles.

Luís Freitas et al. (2021), defined a set of defects that occur in the manufacturing process in PCBs and that affect the productivity and quality of the final product, namely [1]:

- Perpendicularity between the receptacle axis and the test plate [1];
- Precision placement of the receptacle [1];
- Defects that may be present in the needles according to their entire routing process until they are placed by the operators on the PCB (this is where this article focuses) [1];
- The angle and force applied by the technician to the receptacles [1];

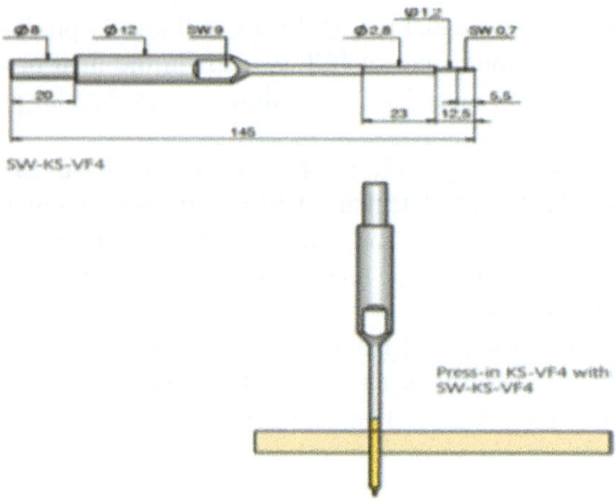

Fig. 2. Needle beds [1].

This work focuses on the process of analyzing the quality of the needles through the implementation of an artificial vision system that will allow a correct insertion of good quality needles in the receptacles, thus avoiding the rejection of PCBs.

To achieve the proposed objectives, the article is organized as follows: Sect. 2 presents a brief description of systems that exist on the market and that use vision systems to detect defects in needles. Section 3 describes how an artificial vision system works and what equipment is used in industrial applications; Sect. 4 presents the vision system that is used for needle quality analysis; Sect. 5 presents the main results, focusing on the most important aspects and also the techniques used to obtain them. Finally, the last section presents the conclusions summarizing the objectives achieved with this solution and the steps to be taken in the future.

2 State of Art

In order to present a conceptual solution to solve a problem related to the quality of needles, this chapter presents in Subsect. 2.1, a description of current systems and in which manufacturing processes are most used in the medical industry.

In Subsect. 2.2, applications with vision systems in the analysis of needle quality in the textile industry are presented.

Finally, in Subsect. 2.3, more applications in other industries with vision systems in the analysis of needle quality will be presented.

2.1 Needle's Vision Systems in Medicine Applications

Balter et al. (2017) presented a paper on how to explained a series of motion tracking experiments using a vision system in conjunction with an ultrasound system and force detection to guide the position and orientation of the needle tip (see Fig. 3). The system enabled automated venipuncture, which provides and combines 3-D NIR and US imaging software, computer vision and image analysis, and a 9-DOF Needle Manipulator in a portable shell. The device operated by mapping the 3-D position of a selected vessel and introducing the needle into the center of the vein based on real-time image guidance and force [2].

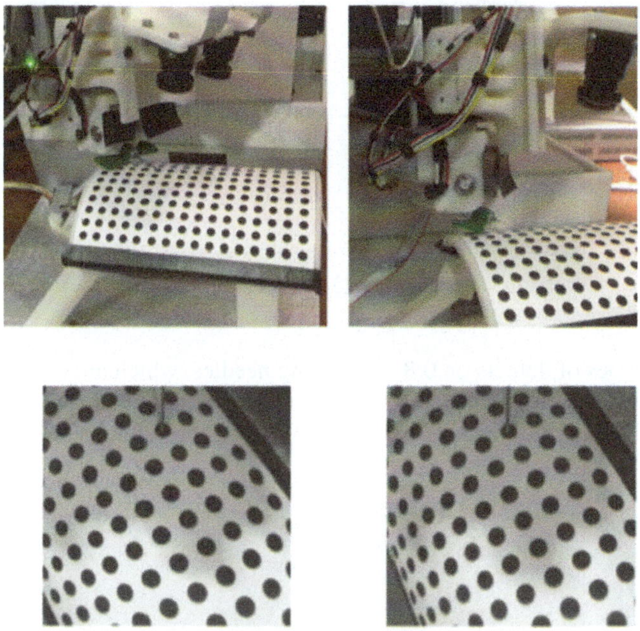

Fig. 3. Needle tip positioning studies [2].

2.2 Needle's Vision Systems in Textile Industry Applications

Zhang, Zhouqiang et al. (2020) presented a paper on how to explained a needle detection system of a hosiery machine based on machine vision (see Fig. 4). A system was then created to detect problems with knitting needles quickly and accurately based on machine vision [3].

The system can detect problems with the knitting needle, and the needle causing the problem can be identified as the beginning of the fabric defect. In image processing, the vertical projection algorithm is introduced into the system, and the image binarization defect is improved [3–5].

2.3 Needle's Vision Systems in Other Applications

Jing Yang et al. (2020) presented a paper explaining how to a stable and accurate real-time tiny parts defect detection system and solve the problems of manual setup of conveyor speed and industrial camera parameters in defect detection for factory products. 0.8 cm darning needles were used as an experimental object.

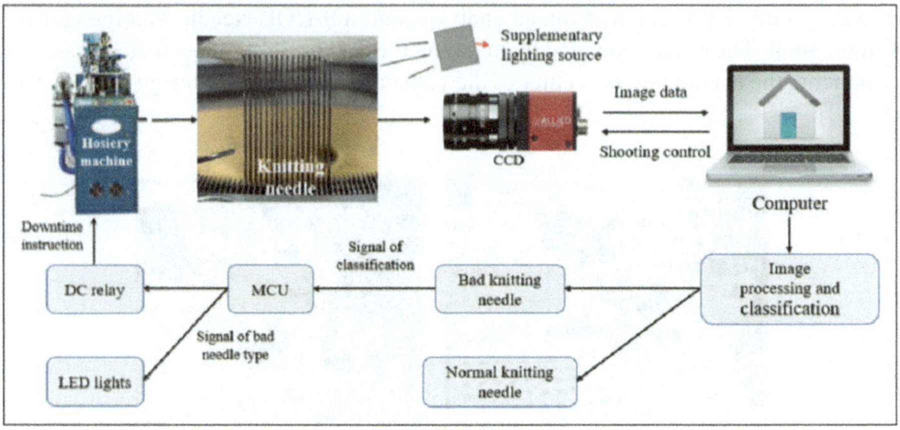

Fig. 4. The principle of system operation [5]

The four types of defects on 0.8 cm darning needles, which are often used in actual production, they are crooked shapes, length size errors, terminal size errors and twist. An industrial vision camera was used in this project to detect defects (see Fig. 5) [6].

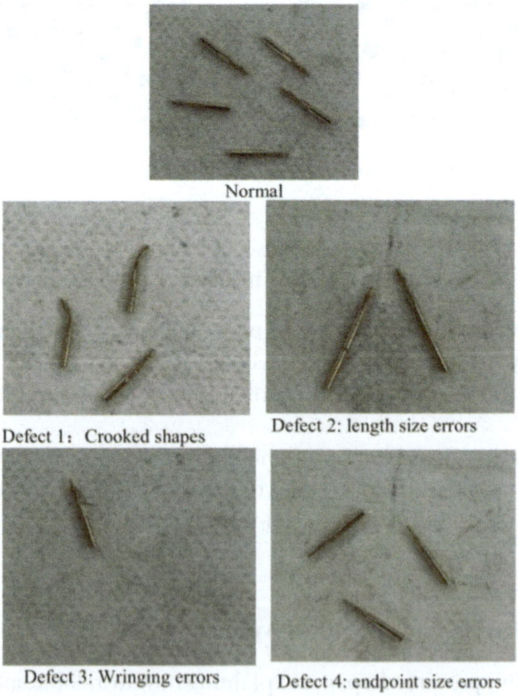

Fig. 5. Autonomous collection of experimental data with needles [6].

At the industrial level, there is currently a gap in industrial PCB manufacturing systems using artificial vision systems to detect defects in needle beds, and it is important to assess whether they can be embedded in PCBs.

3 Vision Systems

3.1 Introduction

Vision systems can be broadly defined as all industrial and non-industrial applications, which through a combination of certain hardware and software components, provide operational directives or inputs to mechanisms/systems based on image capture and interpretation [7].

It is immediately possible to associate terms such as artificial vision and computer vision to these systems, sometimes even wrongly identified as the same, vision systems try to combine all the existing development in these areas in a practical way.

Thus, a vision system needs to take into account all aspects inherent to the inspection process, image capture, lighting, image processing, communication protocols, etc. Therefore, vision systems are defined as the integration engineering of mechanical, optical, electronic and software systems essential for examining objects, products, manufacturing processes and production lines, with the aim of detecting defects, improving their quality, offering more efficient manufacturing processes and production lines and insurance.

They also play a key role in the control of equipment used in manufacturing and are currently used successfully in several high-volume markets, for example, in the control and verification of the correct placement of components on printed circuit boards, plate reading and security (recognition facial, fingerprints and signatures).

The advantages that vision systems guarantee, when well designed and applied, in an industrial environment are mainly:

- Repeatability, capable of providing production lines or manufacturing processes with consistent and reliable inspections for 24 h, 7 days a week without being affected by external constraints, such as fatigue;
- Speed of operation, on a production line a vision system can inspect hundreds and sometimes even thousands of objects/products per minute;
- Accuracy and high inspection detail, in some production lines and manufacturing processes there are details impossible to inspect by the human eye, vision systems when correctly designed (correct camera resolution and optical lens) are able to inspect these same details with facility;
- Non-contact inspection, in the industry where inspection without sample contamination is indispensable, for example, pharmaceutical and food industry, vision systems present themselves as a dominant inspection method;
- Operation in hazardous environments, vision systems can operate in extreme environments, high temperature, toxic or radiation harmful to human beings;
- Eliminates costs associated with the wear of mechanical components of quality control systems and their maintenance time;

- Reduces the need for human labor in production lines and manufacturing processes, increasing safety and profitability;

All these advantages are combined with the objective of improving product quality, reducing waste, increasing profitability and, finally, increasing industry profits. A well-designed industrial vision system requires robustness, reliability, high precision, mechanical stability and adequate cost.

Vision systems are made up of five main components, lens or objective, camera, lighting, image processing unit and communication interface, all of which depend on the object, product or process that will be inspected by the vision system.

3.2 Types and Classification of Vision Systems

An artificial vision system has several components, from the camera that captures an image for inspection to the processing engine itself that renders and communicates the result (see Fig. 6) [7].

The goal of good lighting is to improve and stabilize the image of the object to be analyzed. Thus, lighting correctly can be the key to success. To choose good lighting, both the material of the object to be inspected and the properties of the light source must be taken into account [7].

The functionality of a lens within the artificial vision system is to transfer a clean image of the object that we want to inspect to the sensor element (CCD or CMOS).

There are different valid technologies to capture electromagnetic energy and transform it into an image that can be treated. CCD technology is mostly used although CMOS is gaining ground day by day.

The processor is in charge of capturing the images and applying the appropriate computer treatments to obtain the relevant information in each case [8].

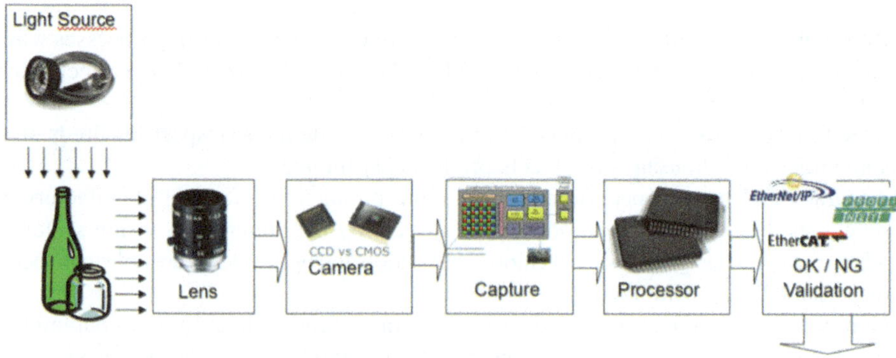

Fig. 6. Schematic diagram of machine vision system.

Below, we make a more detailed analysis from a different types of system visions:

Vision sensors: includes different types of cameras and light sensors at all frequencies, 2D and 3D sensors [9].

Smart cameras: in addition to sensors, they include a processor capable of analyzing images and providing results to make decisions.

Advanced vision systems: These are devices, such as cobots, that integrate artificial vision systems to perform specific tasks [10].

At the level of classification of vision systems, these can be the following:

a) **light source:** bright field, dark field and backlight. Each of these categories refer to the position of the light source in relation to the plane of the object to be illuminated, which results in different angles of incidence.
b) **lens characteristics:** Focal Distance, Maximum aperture, Max sensor size, Distortion, Minimum focus distance, Type of Mount, Iris Adjustment and Focus Adjustment
c) **camera parameters:**

   - Project in which it is inserted and future projects;
   - Technology to be used in the sensor;
   - Camera interface and communication;
   - Drivers and Software (SDK) available;
   - Working Distance (WD) and Field of View (FOV);
   - Minimum sensor resolution;
   - Lighting to which it will be subject;
   - Focus and Depth of Field;
   - Other camera features (Shutter, Color, FPS, etc.).

d) **capture capabilities:** Color or Monochrome, FPS (Frames Per Second), Rolling Shutter vs Global Shutter and Environmental Working Conditions.
e) **processor characteristics:** ARM type processors with interfaces for MIPI/CSI cameras, general purpose processors with x86/x64 architecture in BOX PC format and processors with architectures dedicated to intense graphics processing, such as:

   - GPU (graphics processing unit). NVIDIA and AMD platforms, multi-core architectures that facilitate parallel processing, critical to reducing latencies in image processing.
   - VPU (video processing unit)
   - GPU
   - Jetson

f) **validation algorithms:** Leave-One-Out cross validation, K-Folds cross validation, Stratified K-Folds cross validation, Leave-One-Label-Out cross-validation and Random sampling with replacement cross-validation.

4 Applied Methodologies

4.1 Laboratory Acquisition System

To solve the proposed problem, an artificial vision solution for industrial application was developed. The image of the component to be inspected is taken in an environment suitable for it. The choice of lighting as well as the background color are important to obtain the best results (see Fig. 7).

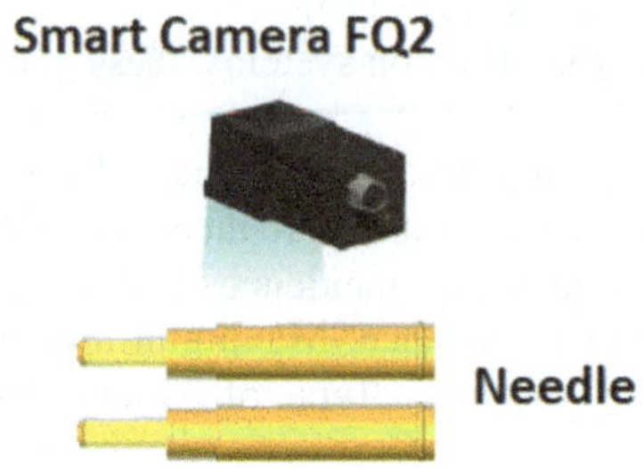

Fig. 7. Laboratory acquisition system

The system consists of a smart camera with a lens and filter attached to it. Also, a lighting system that the camera has is essential to obtain the best possible results. The needles are placed in a strategic position, within the camera's field of vision. The network present in this project is the Ethernet [11, 12].

4.2 Vision System and Software

The FQ vision sensor (see Fig. 8) allows quick and easy real-time parameter adjustment. Eliminates the need to stop the machine for fine tuning and optimization of settings, resulting in zero machine downtime [13].

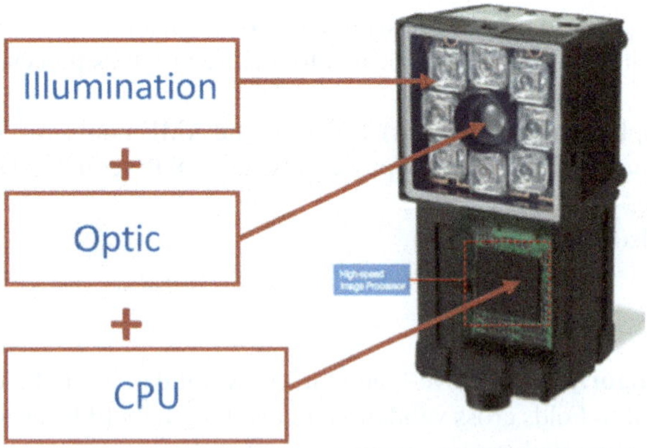

Fig. 8. FQ2 – smart camera from OMRON [13].

The FQ vision sensor combines a high-powered processing unit and true color processing technology that allows fast inspections using color images. The same technology is used in Omron's leading-edge vision sensor model, which is widely used in industry. This smart camera utilizes Omron's high dynamics (HDR) processing technology, improving system dynamics up to 16 times over conventional vision sensors. The result is stable detection of objects that are highly reflectors, even when the placement of parts is not consistent.

The software used for this project was touch finder for PC (see Fig. 9) [13].

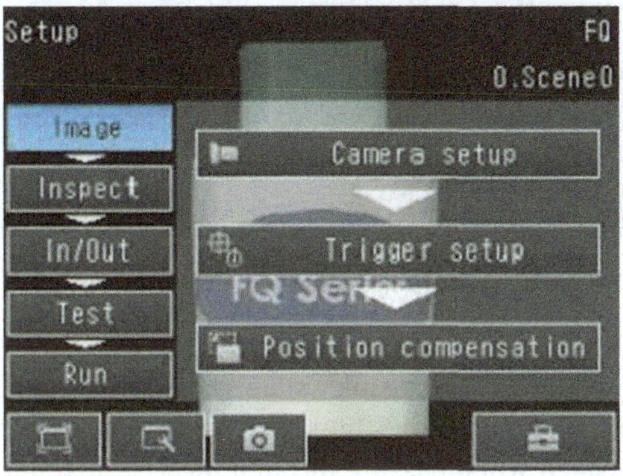

Fig. 9. Software touch finder for FQ2 – smart camera OMRON [13].

4.3 Camera

The choice of the Smart Camera FQ2, for this project, was based in a comparative study of the more important existing ones on the market, presented in Table 1.

Table 1. Advantages and disadvantages of the Smart cameras chosen for the system.

Manufacturer	Model	Advantages	Disadvantages
OMRON	FQ2	Low price; software free	Only Ethernet-IP industrial protocols
COGNEX	In-Sight	More industrial protocols	High price; pay software
KEYENCE	IV3-500MA	More industrial protocols	High price; pay software

5 Analysis and Discussion of Results

5.1 Defects

A new system has been developed that can be used to inspect a variety of different types of needles and geometries fully automatically and can analyze defects in them.

This type of analysis allows a further identification of which needles do not meet the requirements to be used in the manufacture of PCB boards (see Fig. 10) [14].

Fig. 10. Needle reception plate with four sets of holes and four diameters [14].

To carry out the insertion of the needle (see Fig. 11), it must not present defects for a correct insertion. For this, the needles must be arranged in a row so that the quality of the needles can be checked when passing through the vision system.

Fig. 11. Needle type for inspection [14].

5.2 Needle Inspection

To carry out the inspection, the first step will be to obtain images of the needles with the smart camera in a position superior to them. For this we must adjust the camera resolution, through the image exposure time, gain and shutter speed.

After these steps, we get the following image (see Fig. 12).

5.3 Techniques Used to Obtain Images

This project uses the Image filtering technique (Filter Items). We can filter the images that are taken by the Camera to facilitate measurement. This is used in the following cases:

- Cut unnecessary funds so they are not measured;
- To remove noise.

To stably find the edges of marks when other edges have been clearly extracted, gray color filters are used (only for sensors with color cameras). This allows converting an image that was input from a color camera into a monochrome image (see Fig. 13). In addition, we can use the technique called Background Suppression.

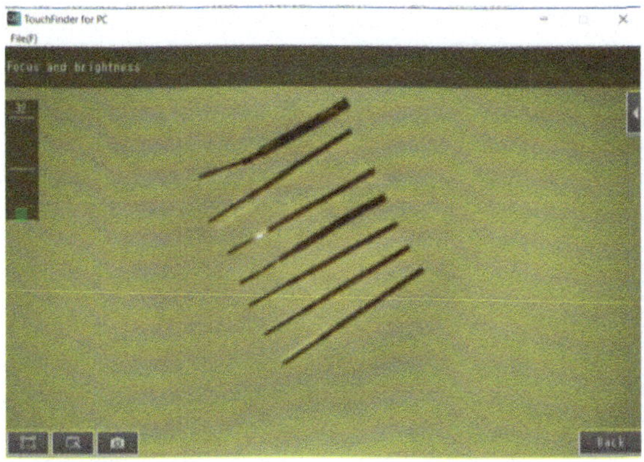

Fig. 12. Needle image with camera adjusts.

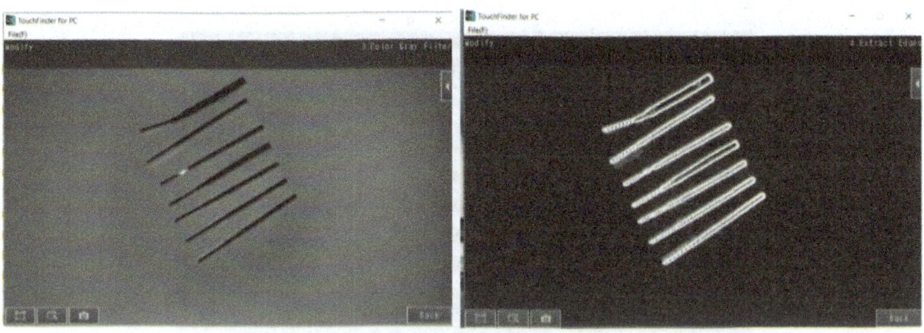

Fig. 13. Needle monochrome image (left) and needle with background removed (right).

5.4 Obtained Results

Was used in the final results, the Edge Position Inspection. This inspection item is used to verify positions needles. For example, it can be used to see if a label or a product is attached at the correct position. In this case Places where the color changes greatly are called edges.

The positions of these needle's edges are measured (see Fig. 14).

In this Teaching function, specifying the measurement region, Character color, Printing type and Correct string, the measurement parameters for OCR are set automatically.

Teaching means to store the region and partial image (see Fig. 15) as reference data for the measurement.

Shape Search II Inspection Item was used in this project (see Fig. 16). This function is for detecting user defined target to estimate target position and pose precisely. The correlation value indicating the degree of similarity, measurement target position, and orientation can be output. In shape search III, edge information is used as features, whereas in a normal search mode, color and texture information are used. It enables

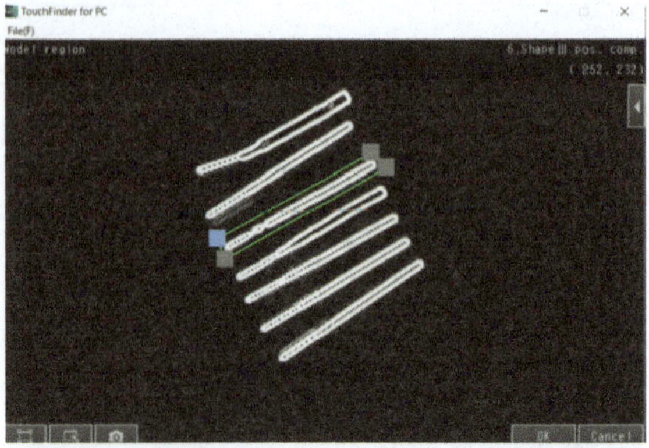

Fig. 14. Needle's edges positions in measurement

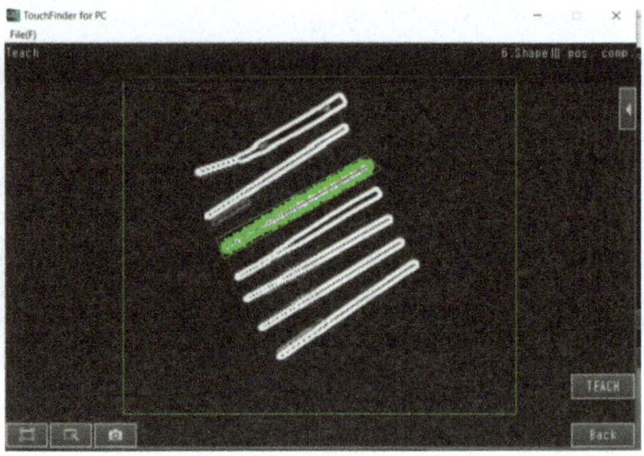

Fig. 15. Needle's measurement region

highly robust and fast detection robust to environmental variations including shadings, reflections, lightings, shape deformations, pose and noises.

Since state-of-the-art object detection algorithm is exploited in shape search II, it can provide much more reliable position and pose estimation with higher speed compared to shape search II (see Fig. 16).

Furthermore, it has much more parameter to tune to support a wider variety of applications.

The needle search area can be expanded by adjusting the function called "Angle range". The OK or NG Judgment is determined by correlation with the image pattern recorded for the search. Therefore, there may be an NG judgment result for a good workpiece if the correlation is low due to the angle being skewed. In this case, to get an OK judgment, increase the Angle range (see Fig. 17).

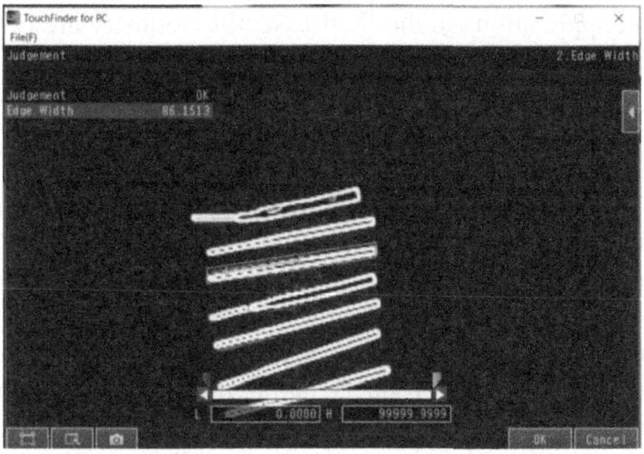

Fig. 16. Shape Search II Inspection in needles.

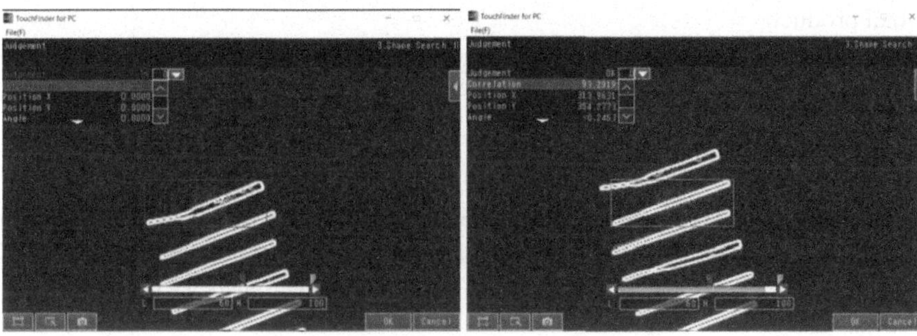

Fig. 17. The OK or NG Judgment in needles.

6 Conclusions and Future Work

The implementation of artificial vision in the analysis of needle positions at the industrial level is currently low compared to other industrial sectors, such as in medicine with its use in conjunction with collaborative robots.

The strengths of this system proposal are: being innovative for this kind of industrial activity, low cost, light weight, easy portability and integration with other existing systems in companies.

The added value part of this study is the fact that, with this proposed system, companies will be able to save money, which comes from the total destruction of the PCB boards. Companies should choose this vision system, as it allows detecting problems in needle beds, and it is important to assess whether they can be embedded in PCBs, taking this process into account.

Currently, the presented technology, in this paper, leaves many promising aspects and allows the analysis of several research and development activities that have been carried out in research institutions and industrial companies.

Machine vision applications in the PCB assembly industry are not yet fully implemented. This is due to the simple fact that they are very specific manufacturing applications that have not yet been investigated and because it is an area that is still under development.

Future use of mechatronic needle routing systems is suggested for better image collection. The inspection on the board that receives the needles must also have a vision system that identifies which ones were correctly inserted and emits an image of those that must be replaced.

In conclusion, the use of artificial vision systems in the electronic PCB assembly industry is beginning to be used and in the coming years it will play an important role in the production and final quality of products, taking into account technological advances and market competitiveness.

The solution here presented makes it possible to eliminate the destruction of PCB boards due to the fact that the needles present defects, which will lead to a better quality in the production process and the consequent reduction of the current quality failures of the final product.

The implementation of this project aims to develop a new technological solution, capable of detecting problems with the needle beds, namely in obtaining the characteristics of each one, such as diameter and length, which are important parameters to consider in this process, among others.

Acknowledgments. This work is co-funded by the European Regional Development Fund (ERDF) through the North Regional Operational Program (NORTE 2020) of the Portugal 2020 Program [Project No. 43922, with acronym "iFixturing"; Funding Reference NORTE-01-0247-FEDER-043922].

References

1. Freitas, L., et al.: Conceptual design of a positioning system for systematic production of needle beds. In: Machado, J., Soares, F., Trojanowska, J., Yildirim, S. (eds.) icieng 2021. LNME, pp. 221–235. Springer, Cham (2022). https://doi.org/10.1007/978-3-030-79168-1_21
2. Balter, M., Chen, A., Maguire, T., Yarmush, M.: Adaptive kinematic control of a robotic venipuncture device based on stereo vision, ultrasound, and force guidance. IEEE Trans. Industr. Electron. **64**, 1626–1635 (2017). https://doi.org/10.1109/TIE.2016.2557306
3. Zhang, L., Li, C., Fan, Y., Zhang, X., Zhao, J.: Physician-friendly tool center point calibration method for robot-assisted puncture surgery. Sensors **21**(2), 366 (2021). https://doi.org/10.3390/s21020366
4. Zhang, Z., et al.: Research on the knitting needle detection system of a hosiery machine based on machine vision. Text. Res. J. **90**, 004051751989917 (2020). https://doi.org/10.1177/0040517519899173
5. Pereira, F., Carvalho, V., Vasconcelos, R., Soares, F.: A review in the use of artificial intelligence in textile industry. In: Machado, J., Soares, F., Trojanowska, J., Yildirim, S. (eds.) icieng 2021. LNME, pp. 377–392. Springer, Cham (2022). https://doi.org/10.1007/978-3-030-79168-1_34
6. Yang, J., Li, S., Wang, Z., Yang, G.: Real-time tiny part defect detection system in manufacturing using deep learning. IEEE Access **7**, 89278–89291 (2019). https://doi.org/10.1109/ACCESS.2019.2925561

7. Li, B.: Research on geometric dimension measurement system of shaft parts based on machine vision. EURASIP J. Image Video Process. **2018**(1), 1–9 (2018). https://doi.org/10.1186/s13640-018-0339-x
8. Golnabi, H., Asadpour, A.: Design and application of industrial machine vision systems. Robot. Comput.-Integr. Manuf. **23**(6), 630–637 (2007). ISSN 0736-5845. https://doi.org/10.1016/j.rcim.2007.02.005
9. Rusenov, B., et al.: Machine vision systems for intelligent quality control of manufacturing processes. In: TECHSYS 2017 (2017)
10. Bhowmik, D., Appiah, K.: Embedded vision systems: a review of the literature. In: Voros, N., Huebner, M., Keramidas, G., Goehringer, D., Antonopoulos, C., Diniz, P.C. (eds.) ARC 2018. LNCS, vol. 10824, pp. 204–216. Springer, Cham (2018). https://doi.org/10.1007/978-3-319-78890-6_17
11. Silva, M., Pereira, F., Soares, F., Leão, C.P., Machado, J., Carvalho, V.: An overview of industrial communication networks. In: Flores, P., Viadero, F. (eds.) New Trends in Mechanism and Machine Science. MMS, vol. 24, pp. 933–940. Springer, Cham (2015). https://doi.org/10.1007/978-3-319-09411-3_97
12. Leão, C.P., Soares, F.O., Machado, J.M., Seabra, E., Rodrigues, H.: Design and development of an industrial network laboratory. Int. J. Emerg. Technol. Learn. **6**(Special Issue 2), 21–26 (2011) https://doi.org/10.3991/ijet.v6iS1.1615
13. OMRON homepage. https://industrial.omron.pt/pt/products/fq2. Accessed 02 Dec 2021
14. I.-T. Solutions. http://www.insidelimits.pt/. Accessed 02 Dec 2021

KAIZEN: An Ancestral Strategy for Operational Improvement: Literature Review and Trends

Manuel F. Suarez-Barraza[1](✉), José A. Miguel-Dávila[2], and Manuel F. Morales Contreras[1,2,3]

[1] International Business Department, Universidad de las Américas Puebla (UDLAP), 72810 Puebla, Mexico
manuel.suarez@udlap.mx
[2] Department of Management, Universidad de León, León, Spain
[3] Universidad Pontificia Comillas, ICADE Business School, Madrid, Spain

Abstract. In 2016 marked the 30th anniversary of the publication of Masaaki Imai's book: "Kaizen. The Key to Japanese Competitive Advantage", so conducting a review of the literature to determine value and trend conclusions became essential for the academic arena. The purpose of this article is to carry out a review of the Kaizen literature, exploring and analyzing the articles presented by the term in the last 30 years (1986–2020). Although clearly from 1986 to 2016 are 30 years, the latest advances in Kaizen were included in the three years that we have been in the second decade of the millennium, that is, until 2019 (33 years) to cover all the possible research in these two decades. Our findings explore quantitative and qualitative arguments and proposition. Therefore, according to the literature we detected five evolutionary stages of the Kaizen concept throughout the year.

Keywords: Kaizen · Literature review · Research trends · Kaizen philosophy · Meta-analysis

1 Introduction

It seems that the world is taking a quite radical turn in this second decade of the 21st century. In recent years, the world's organizations have faced political, social, and economic pressures. Pressures that were increased by the real and latent threat of the global COVID-19 pandemic that has paralyzed practically all people and companies in the world [1]. In the same way, we can add to these pressures terrorist attacks, social changes, excess accessibility to technology, even climate and energy change (hurricanes, earthquakes and other catastrophes). The operations of the companies have faced all this turbulent environment in a context of increased competition, global markets, demanding and impatient customers, an increasingly substantive need for integrated supply chains and of course, new workers known as: "Millennia's (generation Y)" who quickly get bored with the jobs they have [2, 3]. Short-term thinking governs the management of Western companies derived from this context, which generates a strong need to achieve results quickly

A. Hamrol et al. (Eds.): MANUFACTURING 2022, LNME, pp. 154–168, 2022.
https://doi.org/10.1007/978-3-031-00218-2_13

and expeditiously. However, this type of management seriously affects the operations of the company, the quality of the processes and of course, the quality of the products and/or services that are delivered to society. For these reasons, organizations have turned their gaze to quality systems and therefore, to continuous improvement processes, also known as Kaizen (改善) for its Japanese word; all with the purpose of generating a long-term vision, reducing environmental pressures, being more productive, reducing costs, and eliminating MUDA (無 駄), from its operations, the Japanese term for waste or activities that do not add value [4–6].

In this way, the subject seems more practical than academic due to its business roots, different countries and companies around the world have adopted it as a technique to improve their operations; from Japan [7] of course, China [8], United States of America [9, 10], United Kingdom [11], Sweden [12], India [13], Holland [14], Mexico [15], until Australia [16]. Now, from an academic angle since the end of the first decade of the 21st century, it has become a "hot topic" of total quality management [17].

On the other hand, different articles have tried to explain Kaizen as a theoretical construct and not simply as a business-like methodological approach. Several of these attempts in the academic arena have carried out reviews of the literature that attempt to explain them, for example, [15, 18, 19]. In 2018, a Special Issue on the subject was even generated in TQM Journal, which indicates that interest in research in this field continues to grow. In fact, the 11 articles included in this special issue provided greater "light" to the theoretical contribution of Kaizen from applications (case studies) to philosophical and theoretical analyzes of it [20]. However, and despite this, the need to explore the issue from a conceptual evolutionary angle is vital and necessary. The theoretical construction of the Kaizen philosophy is preponderant for its application and implementation in 21st century organizations, due to the specific need to face so many pressures from the external environment (including the global pandemic). Thus, the purpose of this article is to carry out a review of the Kaizen literature, exploring and analyzing the articles presented by the term in the last 30 years (1986–2020). Although clearly from 1986 to 2016 are 30 years, the latest advances in Kaizen were included in the three years that we have been in the second decade of the millennium, that is, until 2019 (33 years) to cover all the possible research in these two decades of the new millennium. Finally, our literature review was carried out by constructing the content of the references in evolutionary stages of the Kaizen philosophy throughout all these years.

The article is structured in four parts. The first part develops the Kaizen philosophy in order to have the theoretical framework that serves as a reference for the review of the literature. The second part explains the research methodology. In the third part, the results of the literature review for decades are presented, providing characteristics and management activities of the content of the articles about Kaizen for each identified evolutionary stage. The article concludes with some final conclusions of the results obtained, in addition to the determination of some topics for future research in Kaizen.

2 Literature Review

2.1 What is Kaizen?

Different definitions of Kaizen have emerged throughout these 30 years of history. Of course, as these authors explain in all these years, misunderstandings, practical language, and even definitions have arisen that confuse Kaizen with other terms such as Lean Manufacturing (Lean), Six Sigma, Continuous Process Improvement and of course, Management by Total Quality (Total Quality Management-TQM for its term in English). Therefore, creating a common language about the topic reduces confusion and contributes to a better understanding of the theoretical approach. Therefore, in this section we try to define Kaizen from its roots in Japanese culture.

On the other hand, some authors have directly defined the Japanese term "Kaizen" and others to the western version of "Continuous Improvement". Both seem synonymous and even by western definition the term Kaizen has also been known as Continuous Improvement. However, for the focus of this research, only those articles that define this approach as "Kaizen" will be specifically taken and only two "classical" authors will be referred to who have defined the term Continuous Improvement from its western angle (see Table 1). In this context, it is important to understand that Kaizen is an ancestral concept despite the fact that Imai [21 p. 23] defines it as "a means to continuously improve personal life, home life, social life and work life. In the workplace, Kaizen means continuous improvement for each of the companies' employees (managers and workers alike), at all times of work". Its roots can probably be traced to the 13th and 14th centuries in medieval Japan of the samurai [22]. For Newitt [23], it is defined as the derivation of two Japanese ideograms (Kanjis): KAI (改) - change and ZEN (善) - virtuous, benevolent, to improve; that combined indicated us: "change to improve" or "the principle of continuous improvement". In this sense, Kaizen begins to be visualized then, as a «philosophy of life», in which the personal, family, social spheres and of course the work aspect can be covered [21, 24]. Due to this tendency to understand Kaizen, as a harmonizing principle of the environment with the values of each individual, other authors even establish its origin in the Confucian philosophy, which establishes a deep respect and harmony for the environment, through a balance between the individual and nature [25]. From what is seen, as a principle or «individual spirit» of Cooperation and Improvement, which rapidly unfolds generating a positive impact on society [26 p. 197]. Other authors have defined it in the literature as follows, which are presented in the following table:

Table 1. Kaizen definitions

Authors	Definitions
Brunet and New [7]	A continuous space of activities in which different actors participate in specific roles to identify and ensure improvements that contribute to corporate goals
Suárez-Barraza [24]	A management philosophy that generates incremental improvements in the work method (or work processes), which allows reducing waste and consequently improving work performance and employee satisfaction
Aoki [8]	Kaizen is the combination of three capacities that lead workers to an operational discipline of self-initiative, inter-functional communication, and the initiative of improvement projects
Macpherson et al. [11]	It is the way Japanese employees see the world. It is a metaphor for the understanding of life, at work, in your company, in your growth in your development, creativity and independence

3 Research Methodology

The research methodology used in this research was the systematic review of the literature. Different authors have successfully worked on this type of methodologies to explain a topic of interest in the business management arena. Some authors have explored the topics of Total Quality Management (TQM); Continuous Improvement [19]; Kaizen [15], therefore, the purpose of the research is based on similar works carried out over the years. In this sense, the work carried out by Sánchez and Blanco [19] is a clear contribution to the literary review of continuous improvement as a theoretical construct in the management arena. Your collection of continuous improvement documents has provided support and sustenance for this research.

In this way, the literature review was carried out on primary information focused at all times on peer-reviewed articles with a blind review of scientific journals and "classic" books that have generated a paradigm of knowledge on the subject of Kaizen. Management articles, popular articles, bachelor's, master's or doctoral thesis, as well as unpublished articles, lectures or textbooks were excluded. The literature review focused at all times on scientific journals of research in operations and production management. Specifically scientific journals on quality management, process improvement and continuous improvement. The articles were identified in prestigious international databases such as: Scopus, Web of Science, ABI Inform Global, Emerald Insight, Business Source Premier, and Elsevier Science Direct. Therefore, the selection criteria were the following:

- The first selection criteria focused on the type of document. In other words, we focus specifically on academic articles in which the keyword in the article title or abstract is: "Kaizen". Although the English translation uses the term "continuous improvement", the search was specifically limited to the term "Kaizen"; The reason is that there are other studies on similar topics in this field for the term of continuous improvement [19]. This study specifically focuses on the Japanese Kaizen concept.

- The second criterion was about the time period. It starts from the paradigmatic book by Masaaki Imai [21] "Kaizen the key to Japan's competitive advantage". Therefore, that year is taken for the expansion of the literature because Imai placed the term of Kaizen in the academic arena of business management and operations. On the other hand, it is known that since the sixties Toyota Motor Corporation already practiced Kaizen in its operational processes developed by Taichi Ohno for solving problems using the PDCA cycle [22], even so, the work focused on of the publication of Imai's book, as this book is considered to have played a crucial role in spreading the concept around the world. Therefore, the analysis of the selected articles focused on the definitions established in Table 1 and on the four synthesis points that characterize Kaizen contained in the article section of the Kaizen literature; which functioned as a theoretical frame of reference.
- The third criterion is about the evolution of the concept. From the pioneering articles of the mid-eighties following the publication of Masaaki Imai's book in 1986 to the latest articles relating Kaizen to environmental sustainability, continuous learning, continuous service improvement, and operational innovation. Literally the 30 years end in 2016. However, as a result of the growing expansion of the literature on this term in 2017, 2018 and 2019, it was decided to include these years to reinforce the analysis of the literature.

Likewise, in order not to leave any relevant article on the subject studied (Kaizen), five main scientific journals in the area of Total Quality Management and Continuous Improvement were reviewed in a detailed and systematic way. These journals were: TQM Journal (Emerald); Total Quality Management and Business Excellence (Taylor & Francis); Business Process Management Journal (Emerald); International Journal of Quality and Service Science (Emerald) and International Journal Operation and Production Management-IJOPM (Emerald). The initial general search in any scientific journal yielded 428 articles using only as a keyword: "Kaizen" in the abstract and in the title of the article (the only two parameters that were strictly considered). Counting on these pre-selected articles, we proceeded to the next phase in which an exhaustive review of each article was carried out in its content (not only in the abstract) in order to maintain the three criteria established for the selection of the same related ones with Kaizen. Subsequently, the search was filtered and refined, focusing on the five selected journals to further refine the articles referring to Kaizen. As a final result of the analysis and filtering of the literature review, 267 articles were obtained, selected under the three criteria indicated above. The rest of the articles (161) were discarded because they did not have the established selection criteria and because the word "Kaizen" was used only as a reference or fashion, but the article in its content did not argue any element of what can be called Kaizen (see Table 1 and the points of synthesis of the philosophy). Articles were also found that indicated "Kaizen" but in their content they related it to another management approach such as: Lean Manufacturing (Lean), Six Sigma or some quality control method. Therefore, it was sought at all times for the final selection that the concept of Kaizen was the heart of the purpose of the article. In other words, the main reason for it, and that the term was not used as "fashion" or a substitute for another managerial approach. Finally, it is important to point out that some scientific articles and seminal books did not contain the keyword "Kaizen" in their title, however their content presents elements to determine that the sources refer to the Kaizen philosophy.

In summary, the applied process of the literature review methodology was the following supported by authors such as Machi and Mcevoy [27]:

1. Determine the literature review based on the purpose of the research.
2. Determine the selection criteria for the articles. Search and explore the literature using the selection criteria.
3. Analyze the groups of topics of the selected articles that are presented in the form of meta-analysis (numerical-quantitative data) and in a qualitative way of patterns of the content of each article.
4. Group into evolutionary phases according to the characteristics of each of the decades since Kaizen was coined.
5. Explain in detail each of the evolutionary phases of Kaizen found in the literature review.

Specifically, the meta-analysis consisted of determining per year the number of articles found with the aforementioned characteristics, in the same way the number of articles per specific journals was determined. From the aspect of content analysis (qualitative) by patterns, the evolutionary stages were determined and for each one of them the articles and those specific articles that determined a differentiator of the decade found were quantified. With this last analysis, a synthesis of each of the decades indicated was generated with the best examples of the Kaizen articles of each evolutionary stage.

4 Quantitative and Qualitative Findings

4.1 Evolutionary Stages of Kaizen from Meta-analysis Perspective

As indicated in the methodology section, the first results of the research are presented from the perspective of meta-analysis. Therefore, the number of articles related to Kaizen was determined from 1986 to 2019. Although 30 years of philosophy (from 1986 to 2016) had been considered since its launch in the management arena in 1986 with the Kaizen book by Masaaki Imai, it was decided to include the years 2017, 2018 and 2019 due to the growing trend that the term presents throughout this second decade of the 21st century. Likewise, the only exception was the article generated by Kodama in 1948 as one of the first published examples of Kaizen from a scientific perspective. The result of the meta-analysis is presented in Table 2 with the number of articles per year and per decade, and the quantitative trends can be observed in the graphs of Figs. 1 and 2.

As can be seen in Fig. 1, the trend for articles began in the second half of the 1980s with only 8 references (including the 1948 reference). During these origins there were few articles on the subject, the vast majority of them were focused on showing examples of the application of Kaizen in manufacturing companies, in addition to some seminal books that seek to explain the Kaizen philosophy [21] and Masao's. Nemoto of the Toyota Production System [28] at the Toyota Gosei company. For the decade of the nineties, the number of references increased notably to 74 of which in 1993 and 1997 were the years in which more articles and books were presented with 11. In 1997 precisely, Masaaki Imai

Table 2. Kaizen papers published by year

Year	Article	Year	Article	Year	Article	Year	Article
1948	1	1990	4	2000	4	2010	13
1986	2	1991	4	2001	7	2011	15
1987	1	1992	6	2002	4	2012	15
1988	1	1993	11	2003	2	2013	7
1989	3	1994	6	2004	3	2014	14
		1995	10	2005	5	2015	17
		1996	7	2006	3	2016	15
		1997	11	2007	8	2017	9
		1998	9	2008	8	2018	17
		1999	6	2009	11	2019	8
TOTAL =	**8**	TOTAL =	**74**	TOTAL =	**55**	TOTAL =	**130**
TOTAL OF ARTICLES SELECTED = 267							

Fig. 1. Graph of total Kaizen articles by year (1986–2019)

[29] publishes his second book where he delves into the concept of Kaizen: "Gemba-Kaizen: a common sense, low cost approach to management (gemba-Kaizen, a low cost approach with common sense to manage)" and explains detailed Kaizen techniques such as gemba management, the 5'S and standardization. Same year that Berger [30] made the first approaches to the guiding principles of Kaizen and began to further define the term. The first decade of 2000 the number dropped to 55 references, probably because the possible "euphoria" of Kaizen had already generated the past and was being replaced

by other approaches such as lean thinking (Lean) or six sigma. Now, by the end of the first decade of the second millennium, specifically in 2009, Kaizen begins to re-emerge with 9 references and in the second decade (2010–2019) it has not yet concluded and reaches 130 more references. Notable rise in it. In fact, it seems like a rebirth of the subject. This result confirms the work carried out by Sánchez and Blanco [19], although their study was focused solely on the term "continuous improvement" in their literature review for articles in the English language worldwide. Now, in this decade (2010–2019) the largest number of Kaizen articles and books of the 33 years measured is presented. Even the average per year for this passing decade was 14.4 references (almost 15 per year), while in the 1980s the average was 1.6 references, less than 2 articles per year.

4.2 Evolutionary Stages of Kaizen (Qualitative Analysis)

As a result of the content analysis of the articles (qualitative) and the patterns found, four evolutionary stages of the term Kaizen emerged in the literature. In addition, a stage 0 that explains the possible origins of philosophy in literature. Each of them is listed below.

- **Stage 0.- Origins of Kaizen**
- **Stage 1.- The Kaizen philosophy (1986–1989)**
- **Stage 2.- Kaizen techniques and tools (1990–1999)**
- **Stage 3.- Kaizen in other sectors (2000–2009)**
- **Stage 4.- Kaizen: Learning, Sustainability and other elements (2010–2019)**

4.2.1 Stage 0. Origins of Kaizen

Sawada [22] based on the meanings of the Kanjis (Japanese ideograms) of Kaizen indicates that both symbols both KAI (改) that can be understood as "the person" that is capable of watering (changing) their internal "I". What waters "the person (Kai)" is his tree, that is, the ZEN (善). Therefore, the person can constantly water the trunk and roots of the tree from him so that they grow strong and healthy. Sustained in the meaning of the Kanjis, Kaizen can be understood as a series of personal principles that make you grow as a person, and that assume that our way of life - be it our work life, social life or family life - deserves to be constantly improved, that is, watering our internal tree.

Its visualization and origin in the roots of Kaizen ZEN Buddhism can be traced back to its arrival in Japan in the 12th and 13th century at the hands of Chinese-trained monks such as Dogen and Eisai. Zen Buddhism can be seen more as a practical religion that seeks to observe and experiment, beyond following figures and dogmas [31]. Therefore, during the XV and XVI centuries the samurai warriors in Japan took it as principles generating the Bushido code or way of the warrior. In this sense, for Ishikawa [32] and Imai [21] himself indicate that ZEN Buddhism is the main influence of the creation of the Kaizen philosophy in Japan. For the virtues of visualizing the present, of experimenting, of growing by yourself, reflecting on your failures and your mistakes and developing more every day. In the fifties and sixties when Toyota Motor Corporation had serious problems in sales and production, Taiichi Ohno turned to the principles of ZEN Buddhism for the creation of Kaizen as a way to improve daily work in the company, in addition to its benchmarking in companies. North American. In fact, today the Kaizen philosophy is a fundamental part of Toyota's management model. For Deming [33] who was invited to deliver a series of total quality management seminars at the same time as Toyota's developments, continuous improvement and constancy of purpose were two crucial elements in the foundation of the Kaizen philosophy (the well-known cycle of Plan, Do, Check and Act - Plan-Do-Check-Act (PDCA for its acronym in English). All this idea of Zen Buddhism, the teachings of Deming and the work done by Ohno were reflected in the first writings of the Kaizen of Kodama in 1948 [34] where it expands on the idea of nutrition of the Japanese as a way of caring for the inner tree and its growth.

4.2.2 Stage 1. The Kaizen Philosophy (1986–1989)

In 1986, Random House publishes Masaaki Imai's book, the "Kaizen, the key of Japanese Competitive Advantage (Kaizen the key to Japanese competitive advantage)" in which the master (sensei) Imai consultant of Toyota and several Japanese companies exposes in a practical and simple way the importance of continuous improvement and standardization in organizations. The terms of maintenance, improvement and innovation of standards revolutionized the way of viewing company management in the mid-eighties. His book, which is a "watershed" of continuous improvement both in the academic and managerial worlds, introduces Kaizen as a philosophy of life and business in which each of the workers, from the CEO to the operators, have to improve your day to day your daily work on a daily basis. Imai's book comes as Japanese automakers were beginning to dominate the North American market and is generating growing concern about Japanese management practices. During this decade, other authors published certain articles that saw Kaizen as a force to motivate people to make improvements in their work processes. For example, Rehder and Smith [35] expose the importance of the improvement in the labor relations of the workers. It was definitely a decade in which Kaizen became known in its definition and its guiding principles are explained in detail by Imai, Nemoto and Matsumura in their books and articles. For Masao Nemoto [28], Kaizen was the driving force behind the quality of Toyota products and he indicates in his book that the active management of people has a lot to do with the motivation to improve on a daily basis. In summary, the evolutionary stage understood as an "Awakening" (see Fig. 2) was characterized by the creation of the bases of the definition of Kaizen, its characteristics and first guiding principles through practical examples in the seminal books of the subject and the first articles developed.

4.2.3 Stage 2. Kaizen Techniques and Tools (1990–1999)

The 1990s represented the evolution of Kaizen as a component of other managerial approaches such as Lean Thinking. James Womack and Daniel Jones' 1990 [36] book "the Machine that Changed the World" summarized five-year research sponsored by the Massachusetts Institute of Technology called: "International Motor Vehicle Program –IMVP- (International Motor Vehicle Program)". Womack and Jones in their book describe the types of production systems in the Western view and in the Eastern view of the world. The result was that the Eastern view described the ability to produce with fewer materials, inventories, people, space, and time. This term was known as "Lean" (slim).

Introduced into academia by the Sloan Management Review article by then-Hyundai CEO John Krafcik [37]. Both authors took the Toyota Production System off the ground in the form of Lean thinking in the academic and practical managerial world. In this sway of concepts, Kaizen went hand in hand with "Lean" thinking to form the elements to establish just-in-time production systems as Toyota implemented it. For this reason, the literature of the Kaizen decade focused on explaining applications and implementations of techniques and tools related to Kaizen and Lean.

Since the Japan Human Relations Association in 1991 [38] they begin to develop Kaizen techniques and their application, for example, Kaizen-Teian where the system of proposals, suggestions or ideas for improvement in organizations is developed. Japanese; going through the MUDA elimination and Kaizen. By the middle of the decade, links with TQM are developed through practical tools and techniques. While Tanner and Roncarti [39] link it with the possibility of obtaining the Shingo Prize for quality in Japan. By the end of the nineties, authors like Berger [30] further delimited the guiding principles of Kaizen, while Sheridan [40]; Melnyk and co-authors [41] and Laraia et al., [10] develop new techniques for applying Kaizen from a Western perspective focused on improving processes in a short time, quickly in weekly periods or fortnightly. These new techniques of philosophy were known as: "Kaizen Blitz, Short-Term Kaizen and Kaizen Events".

4.2.4 Stage 3. Kaizen in Other Sectors (2000–2009)

With the arrival of the new millennium, the expansion of Kaizen begins to other sectors of the industry beyond manufacturing. However, as noted above, the trend during this evolutionary stage was slow and much lower than in the 1990s. Only 55 articles. Some examples from this decade of articles that made new contributions to the Kaizen philosophy were the works of Styhre [25] where he relates Kaizen with ethical dilemmas when empowering employees; Bodek [42] with a methodology that goes beyond Kaizen Blitz, called: "Quick and easy Kaizen". And of course, the work of Brunet and New [7] in Japan.

By the middle of the decade, articles referring to the application of Kaizen began to appear. Emiliani in 2005 [43] proposed the implementation of the philosophy in business school graduate programs. While Suárez-Barraza and colleagues in the public management of Spanish municipalities [44], calling them: "Lean-Kaizen Public Service". Finally, other works that were fundamental at the time and the most cited articles in the decade were those of Farris and colleagues about the critical factors in the Kaizen events in 2009, the transfer of Kaizen to other countries such as Aoki's China.

4.2.5 Stage 4. Kaizen: Learning, Sustainability and Other Elements (2010–2019)

The second decade of the 21st century keeps Kaizen in organizations and in the academic debate. Proof of this is the publication of the Special Issue of TQM Journal in 2018 on the subject of Kaizen. Derived from the resurgence of the academic debate about this philosophy, this stage has been called: "re-birth stage", but in reality these nine years of the decade have become, in addition to the resurgence in expansion in different emerging countries with multiple and varied bonding themes.

Thus we have the Kaizen implementation work in Mexico: Suárez-Barraza and Ramis-Pujol [45] in public service companies. For the maquiladora companies in northern Mexico, the work on the critical factors of the Kaizen events by García and colleagues in 2013 [46]. Chanda in 2012 [47] for the application of Kaizen in Zambia, in South America in Brazil with Raiset, Dos Santos and colleagues in 2016 [48] for the implementation of gemba-Kaizen in a factory, and Ramírez-Alvarado in an exploratory study of Kaizen in Ecuador and Mexico [49]. In Europe, in the Netherlands with the transfer of Kaizen to Dutch companies with Yokozawa and colleagues [14]; and finally, in Asia, in India with the work of applying Kaizen in small manufacturing companies by Ayria Kumar [50] and Prashar [13]. And of course, in China with the research of Zhibin and colleagues to find the Kaizen success enablers (enablers) in companies in that country. Figure 2 summarizes each and every one of the identified evolutionary stages of Kaizen.

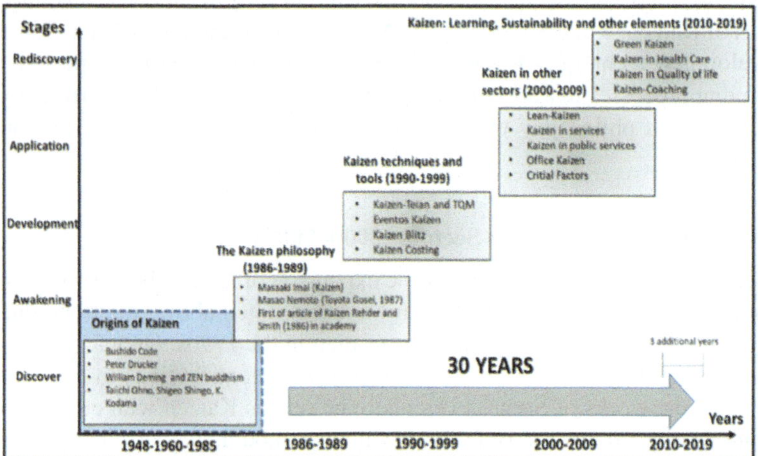

Source: By own

Fig. 2. Evolutionary stage according to the literature review.

5 Final Conclusions and Research Trends

The purpose of this research was to conduct a Kaizen literature review, exploring and analyzing the articles that have been discovered in the last 30 years. As a result of this exploration of the literature, a theoretical scheme emerged (Fig. 2) in which the evolution of the Kaizen concept is shown over the years in four evolutionary stages. Likewise, the quantitative trends in the number of references per year and per decade of the Kaizen philosophy since 1986 were determined. The main contribution of this research is that despite the fact that similar works have been carried out such as those of Sánchez and Blanco of 2014 [19], Singh and Singh [18] and Suárez-Barraza et al., [15] determining the evolutionary stages of awakening, development, application and rebirth of Kaizen (see Fig. 2) with their respective characteristics, contributions and authors of Kaizen differentiate it from past reviews. On the other hand, this exploration focused only on the term Kaizen and not on its western version of continuous improvement [15], which presents an important contribution to the theoretical constitution of Japanese Kaizen. On the other hand, because of the detailed analytical reading of the 267 references, a group of four fundamental learnings emerged from the Kaizen literature review:

1. The definition of Kaizen from its origins in ZEN Buddhism, Toyota, in Imai's first book and with the contributions of all the authors continues to need further theoretical development. However, in these 30 years some common elements of the definition have been built. For example, the best incremental, the emphasis on maintaining and improving standards, eliminating Muda from processes, and of course, the active participation of people in solving operational problems. Every one of the previous elements is understood in the literature as elements that have a common language among the authors of seminal articles and books on the subject. For this reason, this research contributes to laying the foundations for future theoretical constructs to obtain the definition of Kaizen.
2. Despite this, Kaizen has developed techniques and methodologies that have been applied in other countries beyond Japan such as: Kaizen Events, Kaizen Blitz, Kaizen in Public Service among others. Generating an expansion of Kaizen throughout the world. In addition, the Kaizen philosophy has been exported to other sectors beyond manufacturing such as health, education, public services, tourism, among others.
3. Despite the boom in the 1990s, the second decade of the 21st century has shown a re-birth of Kaizen throughout the world. The sample is in the 130 references of these 9 years. However, its literature continues to be very heterogeneous from case studies of implementation in emerging countries and Europe (both in the service and manufacturing sectors), to in-depth research on Kaizen events and their critical factors. Therefore, it is necessary and urgent to develop quantitative research that delimits its guiding principles and critical factors, with a total systemic vision of Kaizen.
4. Like any managerial term, Kaizen is not exempt from criticism, confusion and "fads" in its application and implementation. However, more than three decades of case studies, concept articles, seminal books, and empirical studies have shown that incremental change with this philosophy presents several benefits to companies. Transforming rigid and bureaucratic organizations into flexible and innovative companies with a work culture in which no one takes for granted that things should remain as they are, that is, without improvement and change.
5. According to literature, we can observe that the trend of Kaizen articles and books is moving towards the learning of people, the quality of life of human beings, and the link as topics of the XXI century such as environmental sustainability and physical well-being. However, there are still references in the body of content of articles and books, especially in the last 9 years, which seek to understand the critical factors of Kaizen events, understand Kaizen as a total philosophy from a theoretical and empirical angle, and, therefore, the work carried out on case studies of implementation of Kaizen in the service sector.

The research trends that emerged from this learning were as follows:

1. Deepen the concept of Kaizen as a definition and its historical roots in ZEN Buddhism.
2. State, develop and evaluate the guiding principles of Kaizen in a quantitative way, probably using a model of structural equations.
3. Continue developing case studies that show the application of Kaizen in different areas and sectors. In addition, of course in different countries.
4. To deepen empirically in certain Kaizen techniques to glimpse its impact on organizations. Examples of this are Kaizen Events, 5's, Standardization, Kaizen Teams, Process Improvement, Problem-solving techniques, among others.
5. Kaizen in the health sector to face the COVID-19 pandemic. Its application and its impact on the effectiveness of its techniques in the face of an unexpected flow of contaminated patients.

The research has limitations and as authors we are aware that the methodology used could consider more academic journals including other paradigmatic books on the subject that have been left out of the search carried out. Although six databases were covered, others with more articles covering the topic of Kaizen could be left out. However, both effects can be minimized by the fact that the selected databases have indexed academic journals, of high impact and known worldwide. Another limitation is the language of the selected articles and books. In this research, 98% of the references obtained were English, although references in Spanish, Portuguese and Japanese were also considered (translated in certain sections). However, we are aware that there may be other works in these languages and others in the world. Understanding that English is the preferred academic language.

References

1. Leite, H., Gruber, T., Hodgkinson, I.R.: Flattening the infection curve – understanding the role of telehealth in managing COVID-19. Leadersh. Health Serv. **33**(2), 221–226 (2019)
2. Moyano-Fuentes, J., Bruque-Cámara, S., Einsenberg, J.: La influencia de las redes sociales en la adaptación de los trabajadores al cambio tecnológico. Cuadernos de Economía y Dirección de Empresa **1**(31), 147–170 (2007)
3. Moyano-Fuentes, J., Sacristán-Díaz, M.: Learning on lean: a review of thinking and research. Int. J. Oper. Prod. Manag. **32**(5), 551–582 (2012)
4. Jaca, C., Suárez-Barraza, M.F., Viles-Diez, E., Mateo-Dueñas, R.: Encuesta de sostenibilidad de sistemas de mejora continua: Comparativa de dos comunidades industriales de España y México. Intang. Capital **7**(1), 143–169 (2011)
5. Suárez-Barraza, M.F., Dahlgaard-Park, S., Ramis-Pujol, J., Estrada-Robles, M.: Applying Gemba-Kaizen in a multinational food company: a process innovation framework. Int. J. Qual. Serv. Sci. **4**(1), 27–50 (2012)
6. Oropeza-Vento, M., García-Alcaraz, J., Maldonado-Macías, A., Martínez-Loya, V.: The impact of managerial commitment and Kaizen benefits on companies. J. Manuf. Technol. Manage. **27**(5), 692–712 (2016)
7. Brunet, A.P., New, S.: Kaizen in Japan: An empirical study. Int. J. Oper. Prod. Manag. **23**(11–12), 1426–1446 (2003)
8. Aoki, K.: Transferring Japanese kaizen activities to overseas plants in China. Int. J. Oper. Prod. Manag. **28**(6), 518–539 (2018)

9. Cheser, R.: The effect of Japanese Kaizen on employee motivation in US manufacturing. Int. J. Organ. Anal. **6**(3), 197–217 (1998)
10. Laraia, A.C., Moody, P., Hall, R.: The Kaizen Blitz: Accelerating Breakthroughs in Productivity and Performance. Routledge, Iowa (1999)
11. Macpherson, W.G., Lockhart, J.C., Kavan, H., Iaquinto, A.L.: Kaizen: a Japanese philosophy and system for business excellence. J. Bus. Strategy **36**(5), 3–9 (2015)
12. Jorgensen, F., Boer, H., Gertsen, F.: Jump-starting continuous improvement through self-assessment. Int. J. Oper. Prod. Manag. **23**(10), 1260–1278 (2003)
13. Prashar, A.: Redesigning an assembly line through Lean-Kaizen: an Indian case. TQM J. **26**(5), 475–498 (2014)
14. Yokozawa, K., Steenhuis, H.J.: The influence of national level factors on international kaizen transfer: an exploratory study in the Netherlands. J. Manuf. Technol. Manag. **24**(7), 1051–1075 (2012)
15. Suárez-Barraza, M.F., Ramis-Pujol, J., Kerbache, L.: Thoughts on kaizen and its evolution: three different perspectives and guiding principles. Int. J. Lean Six Sigma **2**(4), 288–308 (2011)
16. Terziovsky, M., Sohal, A.S.: The adoption of continuous improvement and innovation strategies in Australia manufacturing firms. Technovation **20**(10), 539–550 (2000)
17. Awad, M., Shanshal, Y.A.: Utilizing Kaizen process and DFSS methodology for new product development. Int. J. Qual. Reliab. Manag. **34**(3), 378–394 (2017)
18. Singh, J., Singh, H.: Kaizen philosophy: a review of literature. IUP J. Oper. Manag. **19**(7), 902–970 (2009)
19. Sanchez, L., Blanco, B.: Three decades of continuous improvement. Total Qual. Manag. Bus. Excell. **25**(9–10), 986–101 (2014)
20. Suárez-Barraza, M.F., Rodríguez González, F.G., Miguel Dávila, J.-A.: Introduction to the special issue on Kaizen: an ancient operation innovation strategy for organizations of the XXI century. TQM J. **30**(4), 250–254 (2018)
21. Imai, M.: Kaizen: The Key to Japan's Competitive Success. McGraw-Hill, New York (1989)
22. Sawada, N.: The Kaizen at Toyota production system. CHU-SAN-REN Qual. Control Course Nagoya **1**(6), 1–38 (1995)
23. Newitt, D.J.: Beyond BPR & TQM - Managing through processes: Is kaizen enough? In: Industrial Engineering, pp. 1–38. IEE Savoy Place, Institution of Electric Engineers, London (1996)
24. Suárez-Barraza, M.F.: El Kaizen: la filosofía de Mejora Continua e Innovación Incremental detrás de la Administración por Calidad Total, Panorama Editorial, México (2007)
25. Styhre, A.: Kaizen, ethics, and care of the operations: Management after empowerment. J. Manage. Stud. **38**(6), 795–810 (2001)
26. Gondhalekar, S., Babu, S., Godrej. N.: Towards using Kaizen process dynamics: a case study. Int. J. Qual. Reliab. Manag. **12**(9), 192–209 (1995)
27. Machi, L.A., Mcevoy, B.T.: The Literature Review Six Steps to Success. Corwin Press, Thousand Oaks (2009)
28. Nemoto, M.: Total Quality Control for Management: Strategies and Techniques from Toyota and Toyoda Gosei. Prentice Hal (1987)
29. Berger, A.: Continuous improvement and kaizen: standardization and organizational designs. Integr. Manuf. Syst. **8**(2), 110–117 (1997)
30. Imai, M.: Gemba-Kaizen: A Common Sense, Low-Cost Approach to Management. McGraw-Hill, New York (1997)
31. Coll, J.M.: Zen Business: Los beneficios de aplicar la armonía en la empresa. Editorial Profit, Barcelona, Spain (2015)
32. Deming, W.: The New Economics: For Industry, Government, Education. MIT Press, Boston (1993)

33. Kodama, K.: The important points in the improvement of the nutrition of our race. [Waga minzoku eiyō kaizen no zyūten.]". [Kōsei kagaku]. Jpn. J. Public Health **7**(1), 37–39 (1948)
34. Rehder, R.R., Smith, M.M.: Kaizen and the art of labor relations. Person. J. **65**(12), 82–93 (1986)
35. Womack, J., Jones, D.: The Machine That Changed The World. Macmillan, New York (1990)
36. Krafcik, J.: Triumph of the lean production system. Sloan Manag. Rev. **30**(1), 41–52 (1988)
37. de Relaciones, A., del Japón, H.: Kaizen Teian I y II. Productivity Press, Cambridge (1992)
38. Tanner, T., Roncarti, R.: Kaizen leads to breakthroughs in responsiveness—and the Shingo Prize—At Critikon. Glob. Bus. Organ. Excell. **13**(4), 517–531 (1994)
39. Sheridan, J.H.: Kaizen Blitz. Ind. Week **246**(16), 18 (1997)
40. Melnyk, S.A., Calantone, R.J., Montabon, F.L., Smith, R.T.: Short-term action in pursuit of long-term improvements: Introducing Kaizen events. Prod. Invent. Manag. J. **39**(4), 69–76 (1998)
41. Bodek, N.: "Kaizen: KaZam!". T and D **56**(1), 60 (2002)
42. Emiliani, M.L.: Using kaizen to improve graduate business school degree programs. Qual. Assur. Educ. **13**(1), 37–52 (2003)
43. Suárez-Barraza, M.F., Smith, T., Dahlgaard-Park, S.: Lean-Kaizen Public Service: an empirical approach in Spanish local governments. TQM J. **21**(2), 143–167 (2009)
44. Suárez-Barraza, M.F., Ramis-Pujol, J.: Implementation of Lean-Kaizen in the human resource service process: a case study in a Mexican public service organization. J. Manuf. Technol. Manag. **21**(3), 388–410 (2010)
45. García, J.L., Rivera, D.G., Iniesta, A.A.: Critical success factors for Kaizen implementation in manufacturing industries in Mexico. Int. J. Adv. Manuf. Technol. **68**(1–04), 537–545 (2013)
46. Chanda, M.: The study of the relationship between Kaizen practices and operations' performance Improvement in Zambian manufacturing companies. Int. J. Multi-discipl. Res. **1**(119), 1–14 (2017)
47. Raiser, F., Dos Santos, A., Oliveira, C.: Gestão da manufatura enxuta no chão de fábrica. Um estudo de caso do Gemba Kaizen aplicado numa indústria fabricante de máquinas, Espacios **37**(30), 21–28 (2016)
48. Alvarado-Ramírez, K.M., Pumisacho-Álvaro, V.H., Miguel-Davila, J.Á., Suárez Barraza, M.F.: Kaizen, a continuous improvement practice in organizations: a comparative study in companies from Mexico and Ecuador. TQM J. **30**(4), 255–268 (2018). https://doi.org/10.1108/TQM-07-2017-0085
49. Arya, K.A., Choudhary, S.: Assessing the application of Kaizen principles in Indian small-scale industry. Int. J. Lean Six Sigma **6**(4), 369–396 (2015)
50. Ma, J., Lin, Z., Lau, C.K.: Prioritising the enablers for the successful implementation of Kaizen in China: a fuzzy AHP study. Int. J. Qual. Reliab. Manag. **34**(4), 549–568 (2017). https://doi.org/10.1108/IJQRM-12-2015-0173

Employee Motivation According to Amoeba Principles in a Production Company

Marta Grabowska[1](✉), Krzysztof Żywicki[1], and Mateusz Biegała[2]

[1] Poznan University of Technology, Elk, Poland
marta.grabowska@put.poznan.pl
[2] Production Company, Elk, Poland

Abstract. Every production system consists of three elements: people, infrastructure (machines, tools) and work items (raw materials, materials). Without any of them, it is impossible to start production. People are the most important element of the production system because they decide about its creation, goals and tasks [1]. The article discusses the problem of assessing the effectiveness and efficiency of activities related to human resource management and motivation in Amoeba Management system. Motivating and communicating techniques were used as criteria to assess the effectiveness and completeness of this approach. Amoeba principles are described in the first part of the article. Motivating and communicating techniques were used as criteria to assess the effectiveness and completeness of this approach.

Keywords: Production system · Amoeba principles · Management techniques · Motivation · Effectiveness

1 Introduction

In every enterprise whose purpose is to meet human needs before producing goods, there is a production system. According to the definition [2], a production system is a deliberately designed and organized material, energy and information system operated by people and used to manufacture specific products in order to meet the needs of customers. Thanks to the use of system resources, a company can transform streams of input materials into finished products intended for sale. A production subsystem includes the direct processing of materials in technological processes and the assembly of elements necessary to manufacture a given product [3]. The production system is surrounded by other systems and subsystems of the enterprise with which it has specific relationships.

Every production system consists of three elements [4]: people, infrastructure (machines, tools) and work items (raw materials, materials). Without all of them, it is impossible to start production.

A. Hamrol et al. (Eds.): MANUFACTURING 2022, LNME, pp. 169–181, 2022.
https://doi.org/10.1007/978-3-031-00218-2_14

The input and output material stream includes, among others:

- materials
- production equipment
- energy
- staff
- information
- capital

Conversion of input material streams takes place as a result of various processes. The basic activity of a production system is the direct production of products that takes place during the implementation of processes. The links between resources of a production system are called relationships. Connections that enable the effective flow of materials and semi-finished products play an important role in a production system [5].

The measure of acting in accordance with the principle of economy is productivity, which is expressed as the ratio of the obtained effects to the expenditure incurred [1]. Production systems, like enterprises, following the principle of economy should [2, 6]:

- maximize the degree of achievement of the goal with given expenditures, or
- minimize the expenditure of resources to achieve the intended goal

Referring to the second option of minimizing the expenditure it should be noticed that managers have limited influence on resources costs. They can create goals of waste elimination or reduction of resource consumption but they are powerless over government policies that effect raising of prices of basic energy sources, such as electricity and gas. Because of several-fold increase in the prices of these resources, savings are sought n every area of the enterprise. Also in human management, where the issue of efficiency has particular importance [7]. People are the most important element of the production system because they decide about its creation, goals and tasks [4]. If effectiveness and efficiency are a part of all employee activities even related to human resource management and motivation, savings would be easier to gain. It seems that the answer for effectiveness and efficiency in every company area is the Amoeba Management system (AMs) [7, 8]. AMs is analyzed in the article and motivating and communicating techniques are used as criteria to assess the effectiveness and completeness of this approach.

2 Amoeba Management

Amoeba Management is a system that, by keeping independent accounting of small teams (amoebas), implements a management model that engages all employees, allowing for extraordinary flexibility in relation to the market, which engages all employees in reducing costs by increasing their responsibility for their own processes, and shapes employees to the manager's way of thinking [8].

Amoebas are small teams of 5–50 employees that operate as independent companies. The name was proposed by one of Kyocera's employees and is intended to reflect the great flexibility of teams that adapt organizationally to the company's needs [9].

A company that operates in accordance with the principles of Amoeba Management is based on the so-called management accounting, which allows for the accurate and timely implementation of activities in each area of the organization.

Amoeba Management is a production system created by Inamori Kazuo in the 1960s in Japan. Kazuo founded Kyoto Ceramic (now Kyocera), which manufactured ceramics. Ever since the company was founded, Kazuo has felt that for the company's long-term development, it is essential to develop a reliable management philosophy that allows all employees to be fully committed to working for the company without any doubts. The company grew from 28 to 100 employees after five years, and then to 200 and 300. Kazuo took care of everything from product preparation to production to sales, and he knew that he was unable to control the operation of the company in this way. He concluded that he had managed the company successfully up to a point when the company had around 100 employees. His intuition told him to divide the company into small organizations that would have their own accounting. Each organization should be the smallest possible business unit in which the leader would control the balance sheet of the team [10].

Amoeba Management is related to all areas of management, which makes identification of the essence of the method important. Kazuo formulated three basic goals for Amoeba Management [8]:

- Establishment of a balancing system for departments directly connected to the market.
- Shaping employees to the manager's way of thinking.
- Introduction of management in which all employees participate.

Objective 1. - Establishment of a balancing system for individual departments directly connected to the market.

In the precision ceramics industry that Kyocera operated in, orders rarely repeated. In this situation, creating a production cost account a few months after production of a given assortment became unnecessary. For area leaders, it was important to current data in order to be able to make quick decisions on the Gemba. According to the principle of "the highest possible sales, the lowest possible costs", each employee must think about reducing costs, but at the same time act to maximize the value of sales. In Kyocera, the Production Department accounted for half of the employees. If the production leader felt that he had influence on the reduction of production costs, he was certainly not interested in the sales proceeds. Another obstacle in an ever-growing business was finding out where a cost was generated.

Therefore, in order to better control the cost centres and increase employee willingness to increase sales revenues, Kazuo decided to divide the company into the so-called amoebas, or independent entities that prepare their own balance sheet.

One of the few conditions for the functioning of an amoeba is the existence of a clear income and costs incurred to obtain it. Another condition is the business finality of the amoeba, i.e. it must be able to act as an independent business. An example is the Resource Warehouse amoeba. Kazuo feared that separation of such an amoeba was too much fragmentation of the organization, but since there are companies that only trade in a given group of raw materials, in Kyocera the raw material warehouse could exist as an independent amoeba [11]. Another condition for the functioning of amoebas is such a division that allows the implementation of the goals and policy of the entire company [8].

Intra-company Transactions
If the production process is long enough to be divided into individual sub-processes, each of them can be considered a separate amoeba. For example, the raw materials warehouse that sells the production materials has a "sale" on its side, and the production "cost". The Production Department continues to sell its products to the finished products warehouse, and then to the Sales Department. There is intra-company trade in raw materials, semi-finished products and finished products. In this way, each amoeba becomes a unit, such as a small or medium-sized enterprise, and directly feels the principle of "highest sales, lowest costs".

Intra-company transactions also allow for great results from the point of view of quality control because the buyer amoeba does not want a defective product that does not meet quality requirements. In this situation, the selling amoeba is left with a lower-quality semi-finished product and has to incur additional costs related to corrections or disposal.

Objective 2. - Shaping employees to the manager's way of thinking.
From the beginning of Kyoto Ceramic's activity, Kazuo directly managed the work of all departments, visited clients, managed production and research. After dividing a company into amoebas, each leader can easily control the degree of implementation of daily operational activities and lead the team without specialist knowledge or managerial skills. Such a leader becomes aware that he is a manager himself and tries to improve the results of his amoeba and changes his attitude from "If I work, I should be paid" to "I work to ensure salaries for my team members". By entrusting the management of amoeba leaders, Kyocera has successfully educated a multitude of managers.

Objective 3. - Introduction of management in which all employees participate.
In enterprises, there is often conflict between employers and employees. Kazuo, not wanting to waste time and energy resolving internal disputes, concluded that managers and workers must act as family members. A company connected by a family relationship can naturally go in the right direction, even with fierce market competition.

However, this alone may not be enough to resolve conflicts - the management idea and purpose are essential and must convince all employees. Kazuo defined the idea of management as follows: "striving for the material and spiritual happiness of all employees and contributing to the development of humanity and society" [12]. Due to the formulated idea of running the company, all employees were convinced that Kyocera would be constantly developing, and they would have decent salaries [8].

Kazuo, wanting to control all organizations using an independent profit and loss balance sheet, had to simplify it as much as possible so to make it understandable not only for the accountant [13]. The statement of the balance sheet per hour materializes the management principle, which is "highest sales, lowest costs". Based on this statement, each leader can control costs and indicate to team members in which areas they should be reduced. In addition to the income and costs of each amoeba, the balance sheet for the hour also includes added value, which is the difference between them. In each amoeba, the so-called hour indicator.

The leader can calculate how much-added value his amoeba created per hour and react in real-time to threats in the implementation of the production or sales plan [14].

The basis of management is mainly the elimination of Muda (wastes) and offering customers only those products that they need. Each amoeba records its monthly action plan as a specific value in order to be able to control its implementation by comparing the plan with actual costs and revenues. By monthly presenting the added value obtained, the amoeba can react immediately to problems.

The hourly rate does not include the cost of wages and salaries. This is because they cannot be controlled in every amoeba. The amount of remuneration is determined in accordance with the company's policy and the strategy of the HR Department. The amoeba leader would find it difficult to manage them, so he focuses on working time, which is important for productivity [8].

2.1 Creation and Organization of Amoebas

Kyocera has built a management system based on the principle of "function first, organization later". At the beginning of its operation, the company did not have the resources to create separate departments such as accounting or administration, which is why the Control Department was created, which carried out tasks outside the area of research and development and production [15]. Kazuo has had several roles since the founding of the company, ranging from marketer to product development and problem-solving. After these experiences, he concluded that the company must fulfil four functions:

- Marketing and sales - increasing customer satisfaction through order fulfilment and creating added value.
- Research and development - meeting market needs by creating new products.
- Production - producing products that satisfy customers and create creating added value.
- Control - carrying out activities for the effective functioning of the enterprise.

In line with Kyocera's philosophy, every employee should know the role and function of each department.

Kazuo started the division with the Production Department, which has the greatest impact on the company's finances. He created the division according to the existing processes and put a leader at the head of each amoeba [12].

In connection with the development and creation of new products, as well as the creation of a new production plant, the number of amoebas increased in Kyocera, which were created, for example, due to the customer, type of product or plant. In other departments, such as Marketing, there was a similar division.

Between individual processes in the Production Department, a purchase and sale transaction of semi-finished products is possible. An important issue is to fix the bid and ask prices between the amoebas. To do this, start with the selling price at which the customer buys the product and then lower it for subsequent processes. The role played by the manager is important here, as it may happen that one amoeba has a good enough price that it does not have to put any effort into achieving it [16]. Another amoeba, on the other hand, can do everything possible, but cannot meet the price. Such a state is unfair, and the manager should settle disputes between the leaders and set a fair price depending on the costs incurred by the amoeba. The second principle in pricing is that it should be as close as possible to the "rate per hour" in all the individual processes involved in producing the product.

Thanks to internal purchase and sale transactions, the price directly reaches the amoebas, and the production is carried out on its basis. Because production amoebas are independent profit centres, they tend to cut costs to make a profit at a fixed selling price [13]. As a result, employees are aware of the balance sheet, unlike other companies where only the manufacturing costs of products are known.

"Balance sheet per hour" is used to manage the annual and monthly plans and to verify the implementation of tasks by amoeba leaders. Balance sheet data is also used to compile the results of departments and the entire enterprise. As a result, the entire company has one indicator per hour. It is important that the "balance sheet per hour" format is standardized for the entire company in order to be able to quickly recognize problems a given amoeba is struggling with [17].

Business management figures should be available and understandable to all employees, regardless of their position. This is how the managerial awareness of the staff is shaped. The aim of Amoeba Management is management by all employees, therefore the control of results by only managers is insufficient. At the morning briefings in Kyocera, the results of all amoeba activities are presented. As a result, morale in the company is raised and employees feel more responsible for the results of their amoebas [8].

Intra-company Transactions

Until a product is finished and leaves the plant, it has to go through various processes within the company. The amoeba transfer of a product is an intra-company transaction.

Between production processes, the transfer of the product is calculated according to production costs. The costs of an amoeba's process are added to the costs of the previous processes.

When there are transactions between amoebas, it is treated formally and is included in the balance sheet as intra-company sales. The price is determined during negotiations [17] (Fig. 1).

Inaccuracies may occur when amoeba labour costs with a small group of employees are included in the "balance sheet per hour" statement. The cost of working amoebas with high-wage workers will have a low bottom line. This can lead to paying too much attention to labour costs, rather than streamlining the management of the entire amoeba business. In Amoeba Management, an employee is a source of added value, not a cost. In an hourly balance sheet system, the focus should be on the total working time of amoeba members, not the amoeba labour cost [18].

Purchase and sale transactions between amoebas

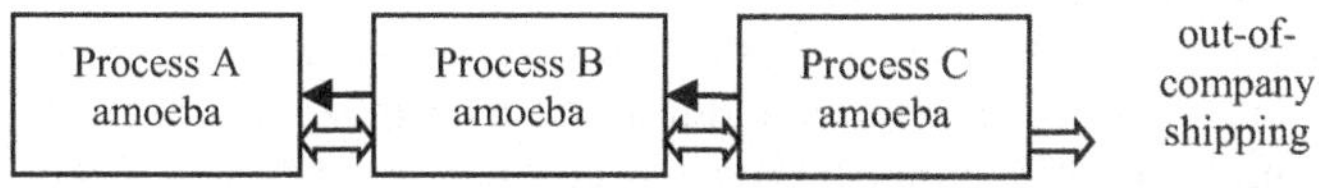

-out-of-company shipping			100	100
-intra-company sales	30	70		100
-intra-company purchase		30	70	100
total production	30	40	80	100

← internal order
⇔ flow of materials, products, cash

Fig. 1. Buy and sell transactions between amoebas [8]

3 Preparation of a Master Plan

The Master Plan is an annual plan that is the basis for monthly forecasts. It should be prepared so as to take into account the strategy of the entire company [19]. To run a business it is important to define goals that must be presented in specific numbers. The Master Plan should be detailed and cover issues such as sales targets, production targets, or employees or equipment. The amoeba leader needs to be ambitious and share this with his employees. According to Kazuo, it is essential to the realization of the Master Plan.

The amoeba leader sets monthly goals that he wants to achieve. The monthly plan should present the results of the previous month and provide corrective actions for any problems that have occurred. When preparing the proposals for monthly plans, the degree of implementation of the Master Plan should be checked. If there is a delay in implementation, actions to catch up with the plan should be presented.

4 Employee Motivation

Knowledge of the terminology and mechanisms of human behaviour in a professional environment is important for managers because they determine the methods of communication, the possibility of delegating powers, and the use of incentives that motivate an employee to act. An employee's willingness to act is a motivation triggered by a specific need. Need means the physiological or psychological lack of something. The best known approach to the analysis of motivation is the hierarchy of needs, in theory and practice, was developed by Abraham Maslow [20]. According to this approach, every human being has lower and higher needs. The lower group includes physiological needs and security, and the higher group includes social, recognition, and self-fulfilment needs.

According to psychologist Douglas McGregor, employee behaviour is influenced, apart from internal needs, by certain attitudes, or rather the manager's assumptions about subordinates. A manager using theory X assumes that employees by nature do not like to work and avoid responsibility and must be forced to carry out tasks because they do not show ambition. When this approach is compared to Maslow's approach, it can be stated that it corresponds to a human being in the pyramid of needs at the second maximum level. Theory Y's attitude assumes that the employee considers the necessity to start a professional job as something natural and not necessary. The consequence of such an approach is taking responsibility for the implementation of tasks, ambition, self-control, creativity, and more [20]. When the characteristics of employee Y are compared with Maslow's pyramid, it can be concluded that the actions of an individual are caused by higher-order needs.

The assumptions regarding motivation were confronted with the employee's expectations towards work and the employer by the psychologist Frederick Herzberg. According to some, his two-factor theory is an alternative to the above considerations, but it can also be concluded that it presents the problem of motivation from the side of applied external stimuli, i.e. it complements the achievements of Maslow and McGregor. According to Herzberg's research, the opposite of "contentment" is not "dissatisfaction" and vice versa. By removing elements that cause dissatisfaction with work, a state of equilibrium will be achieved, or in other words, acceptance; however, the will to act will not exist. According to Herzberg, the factors of the so-called Hygiene measures, the presence of which causes no dissatisfaction, include company policy, interpersonal relations, the availability of tools necessary for the performance of work, and wages. Motivational factors that induce satisfaction and willingness to act include development opportunities, a sense of belonging, additional powers, and division of power [20].

The issues related to influencing employees are aimed not only at achieving a positive atmosphere and job satisfaction, but above all, they should translate into economic effects of the economic organization. In this article, the concept of efficiency is about leadership and employee motivation. Using the relation of effects to expenditure as an indicator, it is possible to substitute the achieved goals of the enterprise in the numerator and the work performed in the denominator [21]. In the context of motivating the team, the effects are [22]:

- task completion - because it is why their team and the process they manage exist. The leader cares about achieving the goal. Failure to do so results in frustration, dissonance, criticism, and ultimately, disintegration of the group.

  The inputs are:
- relationships - both between leaders and team members, and between group members. They can be divided into those that concern the entire team, its morale and sense of pursuing a common goal, as well as relationships with individuals and ways of motivating them.

The key to optimizing the effects-expenditure relationship is commitment and it is assumed that an effective leader is able to acquire, develop and translate them into effects. It is estimated that employees become more involved when [21]:

- they get a job appropriate to their talent, skills, qualifications - that is, the leader is able to find talented people to achieve the intended goal
- they have a good manager - the leader is able to delegate authority to the right people in the organization
- their supervisor is focused on the strengths of the employees - the leader is a humanist, has knowledge about people, and knows how to support and develop employee talents
- a supervisor helps build strong relationships between people - the leader is equally focused on people as entities in the organization, and on achieving the set goals
- the employee is in constant interaction with a leader - the leader knows to keep the employee engaged by being both a coach and mentor
- employees receive information on the progress and results of their work - a leader is able to provide feedback on progress and results
- employees are led towards a positive future - a leader has a vision and is able to communicate it.

5 Motivating Employees vs. Amoeba Principles

The ability to identify the needs of employees and define their attitude towards professional work is crucial for the leader to properly select incentives, communicate tasks, define the powers and scope of control, and more. Contemporary management styles in which the manager first serves higher purposes and leads second is called leadership at a higher level. It is based on four core values: ethics (doing the right thing), interpersonal relationships (fostering trust and respect), success (measured by financial performance), and learning [23–25]. In order to put these principles into practice, techniques and tools are used, e.g. Management by Objectives, results, exceptions, Time Based Management, innovation, conflict, and communication. These techniques are integrated, and some often dominate to adjust the management style to the current situation. In general, it can be assumed that a properly mixed set of practices "Management by xxx" ensures management effectiveness. Motivating and communicating techniques were used as criteria to assess the effectiveness and completeness of this approach (Table 1).

Table 1. Management by xxx techniques vs. Amoeba principles

Management techniques	Principles	Amoeba management
Management by objectives	-joint planning of projects	Yes
	-systematic reviews and self-control	Yes
	-bonus incentives, promotions	Yes

(continued)

Table 1. (*continued*)

Management techniques	Principles	Amoeba management
Management by results	-control and settlement of final results	Yes
	-Profit Centres - Setting a minimum level of profit value to be achieved	Yes
	-cost centres - setting operating cost limits	Yes
	-independence of the organizational unit in the management of the funds granted	Yes
Management by delegation of authority	-organization of operational and routine tasks	Yes
	-responsibility - the scope of delegated powers in writing, proper selection of a person, supervision and control	Yes
	-enriching work and activating employees	Yes
	-appreciating and rewarding employees	Yes
Management by conflict	-activity in the form of "fighting spirit" and striving to weaken undesirable organizational behaviour	?
Management through communication	-assumes that employees want to be partners for management in management processes	Yes
	-prevents the so-called an information void, filled most often by guesswork, rumours and suspicions	Yes
Management by exceptions	-decision-making autonomy of subordinates in standard actions or activities falling within the accepted tolerance ("in plus" - opportunities, "in minus" - threats and crises)	Yes
	-it affects the independence and responsibility as well as trust in the lower management level	Yes

(*continued*)

Table 1. (*continued*)

Management techniques	Principles	Amoeba management
Management through innovation	-identification of problems to be solved, goals	Yes
	-conducting economic analyzes in terms of the profitability of the proposed solutions and the social effects of changes	Yes
	-superiors create conditions conducive to innovation; first of all, they evaluate the creativity and creativity of employees	Yes
Time-based management	-tasks (most often unique) result from the decomposition of the main (and intermediate) goals of the enterprise	Yes
	-assignment of tasks during briefings with employees	Yes
	-at least one-time control of the performance or progress of tasks	Yes
	-wage incentives in the form of piecework pay (per piece, turnover commission) or for the work performed	Yes

Source: Author

6 Conclusion

The implementation of Amoeba Management requires many changes in an enterprise, from the organizational structure, through the financial system, to the organizational culture. It requires time and financial resources, and carries risks. Many studies emphasize the positive impact of the Amoeba principles on the global effectiveness of the company expressed in financial terms, which confirms the success and legitimacy of the concept application. This paper deals with the topic of motivation in the Amoeba concept and the influence of the applied principles on relations and communication between the superior and the employee. It was assumed that depending on the situation, needs, qualifications of employees or teams, and the complexity of the tasks, an appropriate communication technique or a mix of techniques, the so-called Management by xxx is used. The Amoeba principles apply to almost all Management by xxx techniques used here as a criterion for analyzing the effectiveness of motivating. Areas marked as "?" in Table 1 are related to conflict management.

In Amoeba principles, attention is paid to positive relationships, to building a pleasant atmosphere, and honesty. At the same time, a structure is built consisting of independently budgeting units that are profit or cost centres. Attitude towards achieving results can influence competition between amoebas and generate conflict. Management by conflict describes various responses in conflict situations and strategies for dealing with them, including:

- avoiding and procrastinating - denying conflicts; they are "taboo" in the com-pany; as a result, suppress the conflict or postpone it
- enforcement - imposing a solution from the top, from the position of authority
- compromise - encouraging to reach a settlement, on the basis of which the parties to the conflict obtain partial satisfaction of their claims, at the same time giving up some of them
- mediation - the conflict is resolved as a result of the participation and judgment of a mediator (independent third party); the mediator is a person authorized by the company's management board in a situation where the direct superior of the parties to the conflict is not able to resolve the conflict.

The choice of the path depends on many factors, among others the source of the conflict. Appropriate response to disputes can bring many benefits to the company, enhance the fighting spirit, and create an atmosphere for creating creative solutions. The technique of avoiding conflicts or pretending that they are not present can spoil the atmosphere in the team, leading to frustration.

Therefore, on the basis of the observations made, research is planned on conflict situations in companies applying the Amoeba principles. The aim is to analyze the implementation of the Amoeba system and influence the generation of competition and conflicts, to recognize the effective methods of dealing with them, and how these methods are related to the cultural context of the enterprise.

References

1. Grabowska, M., Takala, J.: Assessment of quality management system maturity. In: Hamrol, A., Ciszak, O., Legutko, S., Jurczyk, M. (eds.) Advances in Manufacturing. LNME, pp. 889–898. Springer, Cham (2018). https://doi.org/10.1007/978-3-319-68619-6_86
2. Pająk, E.: Production Management. PWN Publishing House, Warsaw (2011). (in Polish)
3. Fertsch, M.: Optimization Methodology in Adaptive Design of Production Systems. Poznan University of Technology Publishing House (1993). (in Polish)
4. Lachiewicz, S., Matejun, M.: Contemporary Concepts of Production, Logistics and Quality Management. Lodz University of Technology Publishing House (2010). (in Polish)
5. Aqulan, F., Al-Fandi, L.: Prioritizing process improvement initiatives in manufacturing environments. Int. J. Prod. Econ. **196**, 261–268 (2018)
6. Hamrol, A.: A new look at some aspects of maintenance and improvement of production processes. Manag. Prod. Eng. Rev. **9**(1), 34–43 (2018)
7. Jurburg, D., Viles, E., Tanco, M., Mateo, R.: What motivates employees to participate in continuous improvement activities? Total Qual. Manag. Bus. Excell. **28**(13–14), 1469–1488 (2017)
8. Kazuo, I.: Ameba Management. Poznan, Shinsei Consulting (2014). (in Polish)
9. Kanert, M .: How the Japanese healed their airlines. Poznan (2013)
10. Kazuo, I.: Amoeba Management. The Dynamic Management System for Rapid Market Response, CRC Press, Boca Raton (2013)

11. Adler, R.W., Hiromoto, T.: Amoeba management: lessons from Japan's Kyocera. MIT Sloan Manag. Rev. **54**(1), 83–89 (2012)
12. Zeleny, M.: Amoeba: the new generation of selfmanaging human systems. Hum. Syst. Manag. **9**(2), 57–59 (1990)
13. Hamada, K., Monden, Y.: Profit management at Kyocera corporation: the amoeba system. In: Monden, Y., Sakurai, M. (eds.) Japanese Management Accounting – A World Class Approach to Profit Management, pp. 197–210. Productivity Press, Cambridge (1989)
14. Adler, R., Hiromoto, T., Suzuki, H.: Amoeba management and organizational ambidexterity: similarities, differences, and implications for organizational fit and success. Int. J. Product. Perform. Manag. **69**(2), 405–427 (2019). https://doi.org/10.1108/IJPPM-07-2018-0254
15. Yang, F.: The operation mechanism of amoeba's organizational model. Open J. Bus. Manag. **6**, 462–469 (2018). https://doi.org/10.4236/ojbm.2018.62034
16. Kusrini, E., Arifsa, F.S.: Design of amoeba management system in training and consulting services companies: a case study in a training & consulting company in Indonesia. In: 2019 8th International Conference on Industrial Technology and Management (ICITM) (2019). https://doi.org/10.1109/ICITM.2019.8710723
17. Urban, W.: Amoeba management system transformation in the light of organisational change literature. Manag. Prod. Eng. Rev. **8**(1), 16–23 (2017)
18. Kyocera Corp., Consolidated Financial Data (1997–2016)
19. Halawi, L., McCarthy, R., Aronson, J.: Knowledge management and the competitive strategy of the firm. Learn. Organ. **4**(13), 384–397 (2006)
20. Robbins, S.P.: Organizational Behaviour: Concepts, Controversies, Applications. Prentice Hall, Toronto (1998)
21. Dźwigoł-Barosz, M .: Roles of business leaders in modern enterprises. Sci. J. Silesian Univ. Technol. Organ. Manag. **70**, 105–118 (2014). Publishing House of the Silesian University of Technology
22. Armstrong, M., Taylor, S.: Armstrong's Handbook of Human Resource Management Practice. British Library, London (2014)
23. Argote, L., Miron-Spektor, E.: Organizational learning: from experience to knowledge. Organ. Sci. **5**(22), 1123–1137 (2011)
24. Kuo, T.: How to improve organizational performance through learning and knowledge? Int. J. Manpower **5/6**(32), 581–603 (2011)
25. Singh, J.B.: Changes in the contents of corporate codes of ethics: an institutional interpretation. Eur. Bus. Rev. **27**(4), 369–388 (2015)

Increasing Forklift Time Utilization in a Food Equipment Manufacturing Plant with a Kaizen Event

Fernando Gonzalez-Aleu[1(✉)], Daniela Garza-Gutierrez[1], Edgar Marco Aurelio Granda-Gutierrez[1], and Jesus Vazquez-Hernandez[2]

[1] Universidad de Monterrey, 66238 San Pedro Garza Garcia, NL, Mexico
fernando.gonzalezaleu@udem.edu
[2] Instituto Tecnologico y de Estudios Superiores de Monterrey, 64849 Monterrey, NL, Mexico

Abstract. The COVID-19 pandemic impacted several services organizations such as hotels and restaurants. Some studies suggest that 50% of restaurants closed during quarantining days will not reopen in the future, producing a domino effect with their suppliers. Therefore, to survey in this crisis, organizations focused their efforts on reducing waste and/or repurposing their manufacturing operations by introducing new products. The purposes of this paper are (1) to mention an overall strategy followed by food equipment manufacturing plant to address the COVID-19 pandemic challenge; and (2) to describe one of the five Kaizen events conducted to address COVID-19 challenges. This Kaizen focused on increasing forklifts utilization rate from 44% to 80%. The authors followed three steps to achieve this aim: understanding company background, conducting a literature review, and elaborating on a Kaizen event. Practitioners in manufacturing and services organizations could extract several highlights to help them to sort operations problems during crisis time. However, future research still needs to understand the socio-economic impact that COVID-19 pandemic around the world.

Keywords: Optimization models · Material handling · Kaizen · Forklift · Continuous improvement · COVID-19

1 Introduction

In February 2020, after the World Health Organization declared the COVID-19 pandemic, the entire world faced an economic crisis; manufacturing losses, unemployment, poverty rate increase, and closed services organizations [1–3]. Therefore, companies and governments worked together to preserve life, health, and economic stability.

Some of the first publications related to the COVID-19 pandemic impact suggested that about 20% of companies had a contingency plan for events that disrupted the day-to-day operations [4]. Companies worldwide were unprepared for an emergency of this magnitude, forcing them to implement different drastic actions to survey, such as following quarantining regulations, repurposing their manufacturing operations (e.g., producing health equipment, mask, and air ventilators), and increasing process efficiency.

A. Hamrol et al. (Eds.): MANUFACTURING 2022, LNME, pp. 182–193, 2022.
https://doi.org/10.1007/978-3-031-00218-2_15

Unfortunately, manufacturing and services organizations understood late the relevance of having long-term strategies and contingency plans [5, 6].

On the other hand, the COVID-19 pandemic changed society's consumption behaviors and needs, accelerating new business opportunities. Therefore, the purposes of this paper are (1) to mention an overall strategy followed by a food equipment manufacturing plant to address the COVID-19 pandemic challenge; and (2) to describe one of the five Kaizen events conducted. This Kaizen event was focused on increasing the utilization rate of forklifts in FEMP. The authors followed three steps to achieve these aims: conducting a literature review, understanding company background, and elaborating a Kaizen event. Lastly, the authors reflect on this research's theoretical and practitioners' implications in the conclusion section,

2 Literature Review

2.1 Kaizen Events

Kaizen, change for improvement or continuous improvement, is a Japanese concept that Imai [7] defined as improving every day, everyone, and everywhere. To achieve this, Kaize should be understood as a managerial philosophy, as an element of Total Quality Management (TQM), and as a set of improvement tools and methodologies [8]. To achieve this, organizations used different initiatives, methodologies, and frameworks as continuous improvement projects (e.g., Kaizen event, Lean Six Sigma, Six Sigma, and quality improvement projects).

Farris et al., [9 p. 10] defined a Kaizen event as "a focused and structured improvement project, using a dedicated cross-functional team to improve a target work area, with specific goals in a accelerated timeframe." Although a common timeframe for a Kaizen event is usually from one to seven days [10–12], there are days or weeks for planning and data collection activities before the Kaizen event [13, 14].

Several drivers move manufacturing organizations to conduct a Kaizen event, such as improving quality, reducing waste, and increasing productivity [15]. Some achievement in manufacturing organizations that used Kaizen events includes [16–18]: increase production, reduce change over, reduce material use, reduce floor space, reduce distance travel, and others.

Now, it is important to understand the relevant literature available about forklifts.

2.2 Forklifts

The material delivery optimization's interest can be regarded because of a need for a precise delivery schedule to avoid line stoppage and the high cost of transportation. The most used transport vehicles for material delivery at assembly lines are forklifts, two trains, and autonomous guided vehicles [19]. The number of forklifts is always heavily oversized due to transportation fluctuations. Forklift performance depends on traffic in plant and factory design, and finally, because it is still manual transport, the human factor is vulnerable to social disturbances [20]. Therefore, the flexibility offered by forklifts causes a significant loss of efficiency.

Several studies have been conducted to solve transport vehicles utilization from different perspectives, such as capacitated vehicle routing problem with two-dimensional loading constraints [21]; an integrated supply model of manufacturing processes, which included facility location and assignment [22]: and scheduling the delivering of raw material at assembly lines while using the minimum number of vehicles [19].

To meet any efficiency goal is important to study and analyze production and the activities made by the material handling vehicles used in the shop floor, so the variables in the model make it more accurate and closer to the real scenario.

3 Food Equipment Manufacturing Plant Profile

The Food equipment manufacturing plant (FEMP) is a part of a multinational organization. Before the COVID-19 pandemic, FEMP had more than 500 employees, nine product families (mainly for exportation), and used more than one thousand SKUs. Its primary customers were hotels and restaurants.

FEMP's products demand was cycled (see Fig. 1). January and February 2020 showed a better year than 2019; however, with the COVID-19 pandemic, FEMS's products demand decreased 24%.

Fig. 1. 2019 and 2020 FEMP's production

To reduce the financial impact produced by the product demand decrease, FEMP conducted an organizational and operational excellence assessment. First, from the organizational excellence assessment, FEMP identifies a weakness in the lack of strategic planning process. Therefore, FEMP conducted a strategic planning process to repurpose new products in the food sector's manufacturing operations that respond to new customer needs, such as handless machines and small kitchens (ghost kitchens). This strategy is part of a working paper. Second, from the operational excellence assessment, FEMP identified several improvement opportunities. Therefore, FEMP conducted five Kaizen events to improve operations efficiency. This paper describes the Kaizen event worked to increase the forklift utilization rate.

4 Kaizen Event

This Kaizen event consisted of four phases in 14-weeks: preparation (four weeks), analysis (two weeks), implementation (six weeks), and closing (two weeks). During each of these weeks, the Kaizen event team spent two hours per day on Kaizen event activities and the remaining time in the day-to-day work. This is a total of 140 h.

4.1 Kaizen Event Preparation

This phase consisted of four main activities conducted in four weeks: Kaizen team selection, Kaizen event planning, Kaizen team preparation, and data collection. Each of these activities is described as follow:

a) *Kaizen team selection (3 h).* After analyzing the facility operations, an external facilitator recommends eight members: three production supervisors, three forklift operators, one warehouse supervisor, and one external leader. This team was created considering Farris et al., [23] critical success factors to achieve Kaizen event outcomes. The list of team members was shown to the production manager for approval; and his only condition was did not remove employees from the day-to-day operation.
b) *Kaizen event planning (5 h).* Considering the production manager's request, the external facilitator proposes to work in the Kaizen event for the last two hours of the shift every day. This idea was accepted and communicated by the production manager to his employees. Other Kaizen event planning activities include defining a working area, defining stakeholder meetings, defining team member roles, defining team member integration meetings, and identifying other resources needs. These activities were conducted by the external facilitators and the external leader.
c) *Kaizen team preparation (20 h).* Considering the lack of Kaizen team members knowledge in lean and quality analysis tools (e.g., Pareto diagram and cause-effect diagram, etc.), the external leader creates a small lean and quality analysis tools training course. This course was created using information and examples from FEMP day-to-day operations and was taught to other FEMP associates and leaders to develop the same language in the organization and facilitate communication during the meetings.
d) *Data collection (12 h).* Kaizen event leader was responsible for collecting historical data related to warehouse facility, forklift utilization, production planning, product demand, and others. Data that FEMP was not measuring that could impact the progress of this project negatively are inventory, replenishment times, and the consumption rate of raw material in each production line.

Overall, these Kaizen event preparation activities took 40 h (four weeks). Now, the team was ready for a kickoff meeting.

4.2 Kaizen Event Analysis

The Kaizen event analysis phase consisted of two weeks (15 h). During these weeks, the Kaizen team conducted progress meetings with customers and stakeholders to show continuous improvement project progress and solve any time barriers. The main activities conducted by the Kaizen team are summarized as follow:

a) *Kickoff meeting (1 h).* During a kickoff meeting, with the attendance of the plant manager, several stakeholders, and the Kaizen team, the plant manager spoke about the relevance of this continuous improvement project during the lack of production that the FEMP was having in the COVID-19 pandemic days. It was clear the all the attendees the support from the plant manager to this Kaizen.
b) *Understanding FEMP operations (4 h).* The company has eight assembly lines supplied from a central warehouse. Therefore, the company needs a good delivery schedule that enables them to deliver on time, ensuring each line has enough materials to satisfy the demand. To better illustrate the shop floor and the location of each assembly line, the Kaizen team obtained a FEMP layout copy, which was shown in the Kaizen team meeting room.
c) *Demand studies (5 h).* Two demand studies were conducted using data from 2019 and the five months of 2020. First, regarding production volume, it is observed that three out of the eight production lines captured 84% of the production volume. Also, it is observed that the FEMP assembly line L1 represents 48% of the total annual production of the manufacturing company. Second, analyzing demand trends from January 2019 against January 2020, there's a growth in demand of 18%; however, the behavior due to the COVID-19 effect is not repeated along the following months. After March 2020, production decreased 62%. This decrease is related to the temporary closure of main customers, such as restaurants and hotels. Another finding to highlight is the observed seasonality pattern: two straight months' demand rises then falls. The Kaizen team concluded that this seasonality is highly related to customers' needs during summer times.
d) *Forklift's activities studies (5 h).* There are two types of forklifts: standing and seated man or reach trucks and counterbalance forklifts. Forklift's activities study was conducted using a database of 1,361 records of forklift movements. FEMP has nine forklifts of reach trucks and eight counterbalance forklifts. Counterbalance forklifts have 70% of idle time while reach trucks, 58%. Overall, it was detected that they were idle almost 66% of the time. The rent of 17 forklifts represent to FEMP an annual cost of $306,000 USD.

Clearly, there's an opportunity to make better decisions in the forklift management area. Decide how many forklifts are needed in the production site while satisfying the operation.

4.3 Kaizen Event Implementation

This section presents an optimization program based on the real assumptions and conditions observed in the Kaizen event analysis phase. As mentioned earlier, the company is not measuring inventory, replenishment times, or the consumption rate of raw material in each production line; however, the Kaizen team recovered some data through records taken in site production. This information is presented in Table 1.

Table 1. Variables considered

	Type	Demand	Capacity of forklifts	Total time of travel	Idle time	Loading time	Unloading time
Route 1	Raw material	51	1	125.7	281.1	152.9	40.0
Route 2	Raw material	62	1	122.9	157.4	154.8	40.0
Route 3	Raw material	51	1	100.8	312.4	123.5	40.0
Route 4	Raw material	48	1	46.0	476.0	75.0	52.5
Route 5	Raw material	7	1	130.9	0.0	342.9	137.1
Route 6	Finished goods	72	3	82.2	222.5	63.5	90.0
Route 7	Finished goods	91	1	38.8	11.8	73.8	90.0

Note: demand in pallets, capacity in pallets, and time in Seconds.

We also know that the shift lasts 9.5 h, the fork lifters have 30 min to take a break. With this information, the Kaizen team coded a model in python. Object-oriented programming (OOP) was the most suitable way to address the case study. OOP uses objects and classes. A class can be thought of as a blueprint for objects. These can have their attributes (characteristics they possess) and methods (actions they perform). With these classes, functions, and objects, the model would be assigning forklifts to the different routes the company had already established and supplied the demand with the minimum number of vehicles possible.

The main assumptions that were observed, discussed, and established in the company's case were considered while developing the model (see Table 2).

- The capacity of each forklift is limited and known
- Only one pallet could be delivered by cycle in raw material routes, in finished goods routes from one to three pallets
- The demand is known and deterministic
- The shortage is not allowed
- Only one warehouse is available to serve the assembly lines.

Objective function:

$$Minimizez = \sum_{i=1}^{n}\sum_{j=1}^{m} W_{ij} \cdot G_i \tag{1}$$

Table 2. Equations nomenclature

Type	Code	Name
Index	i	For the forklift (i = 1, …, n)
Index	j	For the routes of delivery and pickup (j = 1, …, m)
Data	n	Forklift index
Data	m	Route index
Data	t	Cycle
Data	s	Time of shift
Data	C	Forklift capacity in pallets
Data	C_{ij}	Forklift capacity in pallets to deliver to route j
Data	d_j	Demand of route j
Data	G_i	Cost for leasing forklift
Variables	X_{ij}	Total time for forklift i delivers for route j
Variables	Y_j	Number of pallets delivered in route j
Variables	W_{ij}	Number of forklifts in used for route j

Restrictions:

$$\sum_{i=1}^{n} Y_j \cdot W_{ij} \geq d_j \qquad \forall j = 1, \ldots, m \tag{2}$$

$$Y_j \cdot W_{ij} \leq C_{ij} \qquad \forall j = 1, \ldots, m, \ \forall i = 1, \ldots, n \tag{3}$$

$$\sum_{j=1}^{m} X_{ij} \cdot W_{ij} \leq s \qquad \forall i = 1, \ldots, n \tag{4}$$

$$X_{ij} \geq 0 \tag{5}$$

$$Y_j \geq 0\, and\ integer\, \forall j = 1, \ldots, m \tag{6}$$

$$W_{ij} \geq 0\, and\ integer \tag{7}$$

$$\forall i = 1, \ldots, n\, \forall j = 1, \ldots, m$$

The objective function, given by Eq. (1), minimizes the number of delivery forklifts used in a shift and its cost. Equation (2) ensures that the number of delivered pallets at each route will satisfy the demand. Equation (3) ensures that the total number of pallets delivered in each cycle does not exceed the forklift capacity. Equation (4) guarantees that the total time of delivery of the demand to each route does not exceed the shift time. Finally, Eqs. (5) to (7) state the nonnegative nature of the decision variables.

To validate the model's efficiency and performance, first, a pilot test was carried out to prove that the model works. Data from one week of operation was recovered and then compared with what could have been with the forklift assignment of the model. Due to COVID-19 measures, security and hygiene, they only operate six forklifts. One forklift was assigned to each route except routes five and six (they shared a forklift). It was not possible to complete the demand, so they added three more hours to the shift to satisfy the demand. Operating this way, a 68% of utilization rate was obtained. However, the model made a different forklift assignment, five forklifts, every route has its own forklift except five, six, and seven, with one forklift. The Kaizen team obtained a 91% of utilization rate.

A second pilot test was carried out with one day of operation; the data was entered into the model. With the demand of that specific date, the model assigned six forklifts. Each route has its forklift delivering, except five and six sharing a forklift. This assignment guarantees a 78% of utilization of the forklifts.

A sensitivity analysis was also made with the median with quartiles taken from the real historical data of demand to prove how the model reacts to changes in demand (see Table 3). March 2020 was considered as the sample to obtain the median with quartiles. The analysis was made with the median with quartiles to measure data dispersion to measure variability around the mean. The quartile breaks the data into quarters so that 25% of the measurement is less than the lower quartile, 50% is less than the mean, and 75% is less than the upper quartile.

Table 3. Sensitivity analysis results

Route	Sample		Median		Upper quartile		Lower quartile	
	# Forklifts	Utilization rate (%)	# Forklifts	Utilization rate (%)	# Forklifts	Utilization rate (%)	# Forklifts	Utilization rate (%)
1	1	94	1	91	1	98	1	76
2	1	91	1	87	1	93	1	73
3	1	91	1	87	1	93	1	73
4	1	96	1	92	2	50	1	78
5	0.5	47	0.5	46	0.5	50	0.3	89
6	0.5		0.5		0.5		0.3	
7	1	60	1	58	1	62	0.3	
Average		80		77		75		78

Data recovered from the original records were introduced into the program and the forklift assignation determined that six forklifts were needed to satisfy the demand with an 80% of utilization. With the demand median, the assignment of forklifts to each route stays the same. Six forklifts are considered to supply the demand on time with a utilization rate of 77%. In the assignment for the upper quartile of demand, there are changes to the number of considered forklifts to satisfy the demand, seven forklifts with a 75% utilization. Finally, even fewer forklifts are assigned to each route for the lower quartile. There are considered five forklifts (routes five, six, and seven) share one forklift the rest of the routes are being assigned with one forklift, with a 78% of utilization.

During the last Kaizen implementation meeting, the team presented its findings to the plant manager, customers, and stakeholders. Although the optimization model showed seven forklifts in an upper quartile, the plant manager decided to stay with three more forklifts as a buffer preventing potential problems of maintenance and peak demands. Therefore, this Kaizen event helps FEMP to reduce their annual operation cost by $126,000 USD.

Overall, Kaizen team spent 60 h during the six-week Kaizen implementation phase.

4.4 Kaizen Event Closing

Lastly, during the following two weeks (19 h), the Kaizen work in the Kaizen event documentation, which includes the following deliverables:

a) *Kaizen event file (2 h).* Every analysis, presentation, meeting results, and to-do list were collected file in a physical binder for future reference.
b) *Software application (10 h).* The Python program was converted into a software application to facilitate FEMP personal future forklift analysis
c) *Working instruction (5 h).* A step-by-step working instruction was developed to use the software application.
d) *KPI Follow up (2 h).* Forklift utilization percentage was included in the plant manager monthly meeting review.

Once the plant manager closed the Kaizen event, the Kaizen team was dissolve and each team member continued with their initial roles and responsibilities in FEMP.

5 Discussion

The purposes of this paper were (1) to mention an overall strategy followed by a food equipment manufacturing plant to address the COVID-19 pandemic challenge; and (2) to describe one of the five Kaizen events conducted. First, the authors briefly mentioned the initiatives that FEMP followed up to address the COVID-19 pandemic impact. The FEMP business initiative will be published in the *Operations Management Research* journal special issue in 2022. Second, out of the five Kaizen events conducted to FEMP, the authors considered that this paper presents an unusual application of the Kaizen event, contributing with two insights for continuous improvement practitioners: Kaizen event to mitigate the COVID-19 pandemic impact and Kaizen event with optimization. Ohno [24] suggested the importance of conducting Kaizen during good times; however, because of a lack of contingency plans in FEMP, the organization decided to use Kaizen events as an approach to mitigate the initial impact of the COVID-19 pandemic. On the other hand, instead of conducting several interactions to identify the best solutions, the Kaizen event team decided to apply optimization models to simulate and validate their decisions regarding the reduction of forklifts.

It is important to control the number of forklifts used at assembly lines since material handling costs represent 15% to 70% of a manufactured product [25]. Although this Kaizen event shows significant results, these results should not be generalized, and readers need to consider the following limitations. First, although a traditional Kaizen event takes from one to seven days (one week), the organization could not remove all Kaizen team members from their day-to-day operations during a complete week. Therefore, FEMP leaders found a different Kaizen event structure that adjusted to their needs (14-weeks and 2 h per day). However, this new approach represents a total time of 17.5 days ([14 weeks * five days * two hours]/[8 h per day]); more than twice of the standard Kaizen event structure, impacting negatively on Kaizen event team members interest in this project. Second, FEMP is a young organization with many opportunity areas and several improvement projects are required. During this Kaizen event, the organization also implemented 5S's. Therefore, results are not insolated to the Kaizen event implementation actions; 5S's implementation could impact this Kaizen event results. Third, as the authors mentioned earlier, FEMP does not measure many variables that could make the model even better and more accurate. Many assumptions were made by the company where the available variables used for the model to work (e.g., gasoline consumption per operator). However, the model proposed was successful as the first effort regarding decision-making on forklift management.

Over time, the operational result will become more like the program's results, depending on the adoption of the process by forklift operators. The model can be expanded regarding future work considering other objectives such as inventory on production lines, online raw material replacement times or frequencies, etc. Also, there are techniques utilized to efficiently schedule forklift motion are: dynamic loading, differential evolution heuristic approach, Adaptive Genetic Algorithms (AGA) method, and Evolutionary Algorithm (EA). Dynamic loading involves the AGV handling multiple loads. This will add value to the facility by minimizing the travel time and the number of laps the vehicle travels between jobs [26].

References

1. Kelp, R, Becker, H.: Managing costs in times of COVID-19: focus on transformation, rather than reactive cost cutting in manufacturing industries. OliverWyman (2020). https://www.oliverwyman.com/our-expertise/insights/2020/nov/manufacturing-industries-2030/by-function-new-sources-of-value/managing-costs-in-times-of-covid-19.html. Accessed 23 Aug 2021
2. Barlow, J., Vodenska, I.: Socio-economic impact of the COVID-19 pandemic in the U.S. Entropy **23**(6), 673–695 (2021). https://doi.org/10.3390/e23060673
3. Buheji, M., et al.: The extent of COVID-19 pandemic socio-economic impact on global poverty. a global integrative multidisciplinary review. Am. J. Econ. **10**(4), 213–224 (2020). https://doi.org/10.5923/j.economics.20201004.02
4. Euromonitor: Coronavirus pandemic to rewrite the future of businesses. Euromonitor International (2020). https://www.euromonitor.com/coronavirus-pandemic-to-rewrite-the-future-of-businesses/report. Accessed 23 Aug 2021
5. Simpkins, R.A.: How great leaders avoid disaster: the value of contingency planning. Bus. Strategy Ser. **10**(2), 104–108 (2009). https://doi.org/10.1108/17515630910942241

6. Morrison, M.: Innovation: the lasting legacy of the COVID-19 crisis - industry analysis by Loyale Healthcare. Cision PRWeb (2020). https://www.prweb.com/releases/innovation_the_lasting_legacy_of_the_covid_19_crisis_industry_analysis_by_loyale_healthcare/prweb17051171.htm. Accessed 22 July 2020
7. Imai, M.: Kaizen: The Key to Japan's Competitive Success. McGraw-Hill, New York (1986)
8. Suarez-Barraza, M.F., Ramis-Pujol, J., Kerbache, L.: Thoughts on kaizen and its evolution: three different perspectives and guiding principals. Int. J. Lean Six Sigma **2**(4), 288–308 (2011)
9. Farris, J.A., Van Aken, E.M., Doolen, T.L., Worley, J.: Learning from less successful Kaizen events: a case study. Eng. Manag. J. **20**(3), 10–20 (2008). https://doi.org/10.1080/10429247.2008.11431772
10. Natale, J., Uppal, R., Maggelet, N., Wang, S., Taylor, J., Ogrinc, M.: The impacts of Kaizen event duration on Kaizen success and logistics. In: IIE Annual Conference. Proceedings, pp. 971–979. Institute of Industrial and Systems Engineers (IISE), San Juan, Puerto Rico (2013)
11. Doolen, T.L., Van Aken, E.M., Farris, J.A., Worley, J.M., Huwe, J.: Kaizen events and organizational performance: a field study. Int. J. Product. Perform. Manag. **57**(8), 637–658 (2008). https://doi.org/10.1108/17410400810916062
12. Aleu, G., Flores, F.F., Perez, J., Gonzalez, R., Garza-Reyes, J.A.: Assessing systematic literature review bias: Kaizen events in hospitals case study. In: Proceedings of the 10th International Conference on Industrial Engineering and Operations Management (IEOM), pp. 1–8. IEOM Society, Dubai (2020)
13. Baril, C., Gascon, V., Miller, J., Côté, N.: Use of a discrete-event simulation in a Kaizen event: a case study in healthcare. Eur. J. Oper. Res. **249**(1), 327–339 (2016). https://doi.org/10.1016/j.ejor.2015.08.036
14. Martin, K., Osterling, M.: The Kaizen Event Planner: Achieving Rapid Improvement in Office, Service, and Technical Environments. CRC Press, Boca Raton (2017)
15. Garza-Reyes, J.A., et al.: Deploying Kaizen events in the manufacturing industry: an investigation into managerial factors. Prod. Plan. Control **33**, 1–23 (2020). https://doi.org/10.1080/09537287.2020.1824282
16. Vo, B., Kongar, E., Barraza, M.F.S.: Kaizen event approach: a case study in the packaging industry. Int. J. Product. Perform. Manag. **68**(7), 1343–1372 (2019). https://doi.org/10.1108/IJPPM-07-2018-0282
17. Hanna, S., Deines, T.: Experiences of implementing lean manufacturing through Kaizen events. In: IIE Annual Conference Proceedings, pp. 1–5. Institute of Industrial and Systems Engineers (IISE). Portland, Oregon (2003)
18. Melnyk, S.A., Calantone, R.J., Montabon, F.L., Smith, R.T.: Short-term action in pursuit of long-term improvements: introducing Kaizen events. Prod. Invent. Manag. J. **39**(4), 69–76 (1998)
19. Fathi, M., Syberfeldt, A., Ghobakhloo, M., Eskandari, H.: An optimization model for material supply scheduling at mixed-model assembly lines. Proc. CIRP. **72**, 1258–1263 (2018). https://doi.org/10.1016/j.procir.2018.03.274
20. Bauters, K., Govaert, T., Limère, V., Van Landeghem, H.: Forklift free factory: a simulation model to evaluate different transportation systems in the automotive industry. Int. J. Comput. Aid. Eng. Technol. **7**(2), 238–259 (2015)
21. Wei, L., Zhang, Z., Zhang, D., Lim, A.: A variable neighborhood search for the capacitated vehicle routing problem with two-dimensional loading constraints. Eur. J. Oper. Res. **243**(3), 798–814 (2015). https://doi.org/10.1016/j.ejor.2014.12.048
22. Veres, P., Bányai, T., Illés, B.: Optimization of in-plant production supply with black hole algorithm. Trans Tech Publ.. **261**, 503–508 (2017). https://doi.org/10.4028/www.scientific.net/SSP.261.503

23. Farris, J.A., Van Aken, E.M., Doolen, T.L., Worley, J.: Critical success factors for human resource outcomes in Kaizen events: an empirical study. Int. J. Prod. Econ. **117**(1), 42–65 (2009). https://doi.org/10.1016/j.ijpe.2008.08.051
24. Ohno, T.: Taiichi Ohno's Workplace Management. McGraw Hill, New York (2012)
25. Tompkins, J.A., White, J.A., Bozer, Y.A., Tanchoco, J.M.A.: Facilities Planning. John Wiley & Sons Inc., Hoboken (2003)
26. Abbas, A., Mohamed, T., Hazem, M.: Optimization of warehouse material handling parameters to enhace the efficiency of automated sorting and storage systems. J. Manag. Eng. Integr. **11**(1), 72–84 (2018)

Barriers to Digital Transformation in Asset Management

Damjan Maletič(✉) and Matjaž Maletič

Faculty of Organizational Sciences, University of Maribor, Kranj, Slovenia
damjan.maletic@um.si

Abstract. The digital age poses new challenges to the field of asset management. The main purpose of the study is to identify the barriers to digital transformation in asset management. A study was conducted with 15 experts (including university researchers and industry experts) to explore, validate, and classify barriers affecting digital transformation in asset management. As a result, 20 barriers were obtained as a result of the study. Barriers to digital transformation in asset management identified included, misunderstanding the strategic importance of asset management, no clear vision/strategy, existing mindset/thinking and culture, inadequate asset management system, lack of understanding of digital trends, lack of employee knowledge and skills, and misalignment of business and asset management objectives. Outcomes of this study can be helpful for organizations on their way to digital transformation of the asset management.

Keywords: Asset management · Digital transformation · Barriers

1 Introduction and Literature Review

Today, companies are facing the evolution of fourth industrial revolution, also referred to as Industry 4.0. Industry 4.0 is the digital transformation of manufacturing and related industries and value creation processes. The ability to have systems with greater connectivity, more information and, at the same time, greater flexibility, allows companies to have a clearer view of their processes and, consequently, to improve their performance outcome [1]. Crespo Marquez et al. [1] emphasized that digital transformation is no longer just about making processes more efficient, but about creating more sustainable and profitable customer relationships by fulfilling customers' needs more efficiently. As such, digitalization is creating new opportunities for asset management through the emergence of Cyber Physical Systems (CPS). In this regard, digital twins can be used by companies to determine how their physical assets perform or to monitor the performance of their assets. Intelligent asset management systems are becoming more prominent and are directly related to building new capabilities needed to manage and process data and information [2]. Therefore, the availability of industrial IoT, new technologies such as predictive maintenance analytics combined with Big Data, and digital twin simulations are driving the growth of the intelligent assets management (IAM) platforms market

A. Hamrol et al. (Eds.): MANUFACTURING 2022, LNME, pp. 194–202, 2022.
https://doi.org/10.1007/978-3-031-00218-2_16

[3]. Technological innovation is opening the door to a whole new world of asset management. The Internet of Things (IoT) as a key enabler of Industry 4.0 as well as Big Data, cloud computing, mobile networks, virtual reality, digital twins, building information modeling (BIM), real-time monitoring of physical assets are some of the trends currently making their way into the asset management world. The digital age poses new challenges to businesses and is enabled by communication between people, machines and resources [4, 5]. Nonetheless, introducing digital approaches in asset management is rather difficult. It brings changes in the infrastructure as regards technology as well as changes in the processes of asset management. Therefore, it is important to understand the challenges and to recognize potential solutions is vital for an efficient transition to IAM.

The ISO 55000 standard provides a good framework for companies in order to understand different aspects of asset management system [6]. Assets, according to the asset management standard ISO 55000, are items, things and entities that have value or potential value to the organization [6]. Therefore, companies try to maximize the value of their assets by investing efficiently in asset management to get a better return for their company [7]. This can be done by focusing on costs, risks and performance and optimizing them throughout the life cycle of the asset [8]. Physical/engineering assets, are important for creating tangible value for company in organizational settings such as manufacturing, energy and water supply, construction, mining, transportation, and several other areas of industry [9]. For successful implementation of the asset management it is important to understand the needs of the organization. In addition, aligning the objectives of the organization and asset management is critical to delivering quality products safely, cost-effectively, and on time [10]. As such, proper asset management strategy and plan is vital for creating value from assets and achieving business goals.

The introduction of Industry 4.0 enables companies to collect and process a large amount of data throughout entire lifecycle phases of an asset. It therefore offers the opportunity to digitize and automate interactions between different stakeholders, reducing errors and enhance the performance of the asset management processes. For example, Industry 4.0 can provide insights into how to improve the asset maintenance by means of predictive analytics. In addition, the large amount of data on assets and asset management also provides opportunities for the future development of asset management using new technologies. Digital transformation in asset management should therefore ensure that the right business information and operational technology data is available at the right time, throughout the system, and throughout the life of the assets. A prominent example is IAM, representing a shift from traditional preventive maintenance to new condition-based and predictive maintenance approaches. There is also a growing shift towards prescriptive approaches, enabling companies to make actionable recommendations for assets (e.g. decisions regarding asset maintenance, operations and other lifecycle management activities).

Asset-intensive companies need to understand and operate within the changing context. The successful implementation of an asset management strategy must be fully coordinated with the digital strategy of the organization, and both resulting from the strategic objectives of the company [11]. They need to manage their adaptation to change in a way that better fits their strategic priorities. So, a digital strategy in asset management is an important requirement for data-driven decision making, which is indeed an important prerequisite for asset management and an important contributor to efficient and effective asset management system.

On the one hand, there is still a great deal of uncertainty among manufacturers about what the implementation of Industry 4.0 really requires of them - and many are still finding it difficult to even get started. On the other hand, most technology providers have converted their portfolios to Industry 4.0 relatively quickly [12]. Adopting new technologies can help companies be more responsive in the marketplace. Accordingly, companies should also consider these aspects when building an asset management system [13]. Very few leaders have the commitment and fortitude required to bring about the kind of long-term change needed to equip organizations for the digital future. This paper aims to identify the underlying factors that hinder or stop digital transformation in asset management. In this regard, this paper contributes to the literature and to practice by providing answer to what experts consider to be the main barriers to digital transformation in asset management.

2 Methodology

In the absence of extensive empirical data, this study used a structured approach to elicit expert insight on the barriers to digital transformation in asset management. This paper is based on a study among experts from Slovenian manufacturing companies, consultancies and universities. Study was based on a Delphi method guideline. Delphi method is used as a group technique aiming to achieve the reliable consensus by surveying a panel of experts [14]. As such, Delphi method is useful and has been widely used in many areas of management. However, some challenges arise in implementing this method, such as selecting panel of experts, designing the survey questions, maintaining the number of panel of experts and their commitment, and achieving a satisfactory level of agreement [15]. In this paper, we divided the process into three blocks: (1) sampling of experts, (2) literature review and brainstorming, (3) narrowing down alternatives and ranking (see Fig. 1).

Since the number of participants (experts) is limited, they must have adequate knowledge and experience of the topic under discussion so that they can represent a variety of perspectives. The population from which the expert panel were drawn comprised academics and expert practitioners working on the field of asset management as well as representative of asset management solution providers. Many of the experts involved are part of various European and international committees and societies in the field of asset management. In total, 15 experts were included in the research.

Traditionally, similar studies begin with an open-ended questionnaire. However, as a common and well-accepted change in format, questionnaire can also be designed based on a literature review. Accordingly, the questionnaire was developed from the literature review. Summarized results are presented in Table 1.

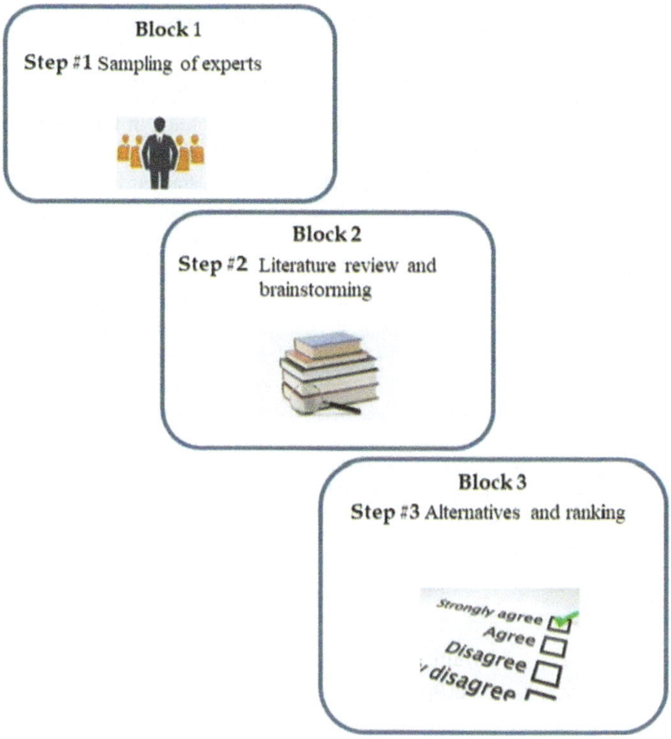

Fig. 1. Research approach

As mentioned above, a questionnaire was developed based on literature review. The survey was distributed to the selected group of experts via the online platform. All terminologies, factors and types of information were explained and defined in the survey. Experts were contacted individually if further explanation was needed or they encountered technical difficulties. A deadline of four weeks was set for the collection of responses. In addition to the evaluation of the questionnaire, experts had an option to add additional barriers, which were not identified based on the literature review.

3 Analysis and Discussion

15 experts agreed to participate in the study. Barriers identified from the literature review were rated on a five-level Likert scale by the panel of experts. Items which have mean value greater than 3 were obtained. Prior studies have also used mean value as a selection criteria [21]. During the research process experts have added additional barriers alongside those identified from the literature. The final result is presented in Table 2.

Table 1. Identified barriers in management literature - summarized literature review

Authors	Research method/focus	Barriers
Stentoft et al. (2021) [16]	Questionnaire-survey (drivers and barriers for Industry 4.0 readiness)	Lack of knowledge about the new digital technologies, lack of standards, more focus on operation at the expense of developing the company, lack of data protection (cybersecurity), lack of qualified employee, lack of employee readiness, requires continued education of employees, lack of understanding of the strategic importance of the new digital technologies, lack of understanding of the interplay between technology and human beings, too few financial resources, too few human resources (manpower)
Marzena et al. (2020) [17]	Multiple case studies (barriers, success factors and leading practices)	Complexity of the logistics system and underlying processes, lack of resources including skilled resources, technology adoption, resistance to change and data protection
Schroeder et al. (2019) [18]	Interviews, focus group, Delphi-based inquire (conceptualization of the industry 4.0 context)	Inhibiting culture, lack of digital exchange standards, business value uncertainty, resource limitations
Peillon and Dubruc (2019) [19]	Literature review and case studies (barriers to digital servitization)	Digital transformation and servitization are not considered as strategic goals
Trappey et al. (2017) [20]	A review of essential standards and patent landscapes (for the Internet of Things)	Lack of standards

Table 2. Results of the research

Barriers to digital transformation in asset management	Mean value
Misunderstanding the strategic importance of asset management	**4.53**
No clear vision/strategy	**4.47**
Existing mindset/thinking and culture	**4.40**
Inadequate asset management system	**4.33**
Lack of understanding of digital trends	**4.33**
Lack of knowledge and skills of employees	**4.33**
Misalignment of business and asset management objectives	**4.07**
Lack of management support	**4.00**
Inadequate hierarchy of physical assets	**3.80**
Lack of understanding of the organisation's key success factors	**3.73**
Lack of understanding and knowledge of processes	**3.60**
Insufficient human resources	**3.47**
Lack of innovation potential	**3.47**
Current IT structure	**3.33**
Inflexibility of processes	**3.27**
Lack of employee readiness	**3.14**
Insufficient financial resources	**3.13**
Insufficient data protection (cyber security)	**3.20**
Dependence on other technologies	**3.13**
Rigidity of regulatory bodies	**3.06**

To the best of our knowledge, there is no current stream in the asset management literature that deals with barriers to digital transformation in asset management. Addressing this gap, we reviewed current general management literature on the topic and combined it with findings from the case analysis. The alignment of what literature reports and experts perceive resulted in the list of barriers to digital transformation in asset management.

Technological innovation is driving organizations to change the traditional way of working in asset management through digitization, with the expectation that this change will help manage risk and improve costs and performance. However, companies are likely to face obstacles on this journey. Findings of our study are consistent with prior research e.g. [16, 22] indicating that barriers related to legislation, management and workforce are affecting the digital transformation process. However, in addition what can one found in management literature, our study highlights the importance of strategic view on asset management. Moreover, the findings suggest that in order to be successful in digital transformation organization should establish asset management system. This is in a line of studies (e.g. [23]) highlighting that organizations should recognize the need for a strategic approach to asset management. It is also well recognized that asset management system supports organization in achieving better performance outcome [8, 24, 25]. Further, strategy is in literature often consider as an important driver of digital transformation [16]. Likewise, in meeting organizational objectives, the asset management strategy provides alignment between the asset management processes and the strategic aims of the business [7]. Since digital transformation should be guided by the broader business strategy, it is essential that asset management strategy reflects the company's objective regarding digital transformation. Hence, asset management objectives included in the strategic asset management plan (SAMP) must be aligned to and consistent with the organizational objectives [6]. This implies that inexistent SAMP in organization represent an obstacle to digital transformation in asset management. SAMP is also the starting point for the development of asset management strategies, objectives and plans that lead to an optimal combination of asset life cycle activities - depending on criticality, condition, performance and risk level. SAMP is therefore essential in setting up the base for introducing new strategies and technologies. Although the asset management industry can benefit from advances in digital technology, this task would be rather difficult to implement without a clear strategy and focus on establishing an asset management system and processes. The primary function of asset management is to preserve the value of assets for the achievement of business objectives [7]. However, misalignment of business and asset management objectives prevents the organization to realize value from assets in the desired level. This also applies to digital transformation as well. Often lacking is critical understanding of how digital impacts the business, and how to effectively plan and deploy the critical skills needed. In such cases, it can happen very quickly that asset management objectives concerning the digital transformation will not reflect the strategic objectives. Digital doesn't just mean 'remote'. For example, machine learning can be used to analyze text information in maintenance notifications to propose suitable failure modes. As such, understanding digital trends, developing digital skills and knowledge is of key importance. Asset management requires competences that meet the requirements of knowledge, skills, experience, behavior, attitudes and attributes related to asset management [26]. New asset management skills are needed for introducing digital transformation. So, the top management needs to create an environment in the organization where asset managers can become excellent professionals who are able to make the right decisions based on and with the help of data analysis. A lack of knowledge and support from top management is therefore an obstacle on the way to digitizing asset management.

In addition, our study shows that the current IT structure, data protection and insufficient financial resources are also barriers that hinder digital transformation in asset management. Prior literature e.g. [27] has shown that inadequate IT structures, lack of technical skills, inadequate business processes, and high implementation risks and costs are often mentioned barriers by companies.

4 Conclusion

Nowadays, companies can no longer escape the opportunities offered by digitalization. Asset management is no exception. Integrating digital technologies into area of asset management fundamentally changes the way you work and deliver value from assets.

However, many barriers can arise when a company decides to embark on the digital transformation journey. In this study 20 barriers related to management, workforce legislation and other aspects were identified. This work can be useful for any organization interested in digital transformation in asset management.

References

1. Martínez-Galán, P., Crespo, A., de la Fuente, A.: A new model to compare intelligent asset management platforms (IAMP). IFAC-PapersOnLine **53**, 13–18 (2020)
2. Crespo Márquez, A., Macchi, M., Parlikad, A.K.: Fundamental concepts and framework. In: Crespo Márquez, A., Macchi, M., Parlikad, A.K. (eds.) Value Based and Intelligent Asset Management, pp 3–38. Springer International Publishing, Cham (2020). https://doi.org/10.1007/978-3-030-20704-5_1
3. Crespo Marquez, A., Gomez Fernandez, J.F., Martínez-Galán Fernández, P., Guillen Lopez, A.: Maintenance management through Intelligent Asset Management Platforms (IAMP). Emerging factors key impact areas and data model. Energies **13**, 3762 (2020)
4. Kagermann, H.: Change through digitization—value creation in the age of Industry 4.0. In: Albach, H., Meffert, H., Pinkwart, A., Reichwald, R. (eds.) Management of Permanent Change, pp. 23–45. Springer, Wiesbaden (2015). https://doi.org/10.1007/978-3-658-05014-6_2
5. Jasiulewicz-Kaczmarek, M., Legutko, S., Kluk, P.: Maintenance 4.0 technologies – new opportunities for sustainability driven maintenance. Manage. Prod. Eng. Rev. **11**(2). 74–87 (2020)
6. ISO 55000 Asset management—overview, principles and terminology. International Organization for Standardization (ISO), Geneva, Switzerland (2014)
7. Lima, E.S., McMahon, P., Costa, A.P.C.S.: Establishing the relationship between asset management and business performance. Int. J. Prod. Econ. **232**, 107937 (2021)
8. Maletič, D., Maletič, M., Al-Najjar, B., Gomišček, B.: An analysis of physical asset management core practices and their influence on operational performance. Sustainability **12**, 9097 (2020)
9. de Almeida, N.M., Vieira, J., Silva, J.G., e Castro, C.: The impact of asset management development programs in infrastructure organizations. In: Rodrigues, H., Gaspar, F., Fernandes, P., Mateus, A. (eds.) Sustainability and Automation in Smart Constructions. ASTI, pp. 247–258. Springer, Cham (2021). https://doi.org/10.1007/978-3-030-35533-3_29
10. Chattopadhyay, G.: Asset management journey for realising value from assets. In: Misra, K.B. (ed.) Handbook of Advanced Performability Engineering, pp. 429–450. Springer, Cham (2021). https://doi.org/10.1007/978-3-030-55732-4_19
11. Trindade, M., Almeida, N.: The impact of digitalisation in asset-intensive organisations. Netw. Ind. Q. **20**, 4 (2018)
12. Industry 4.0 after the initial hype: Where manufacturers are finding value and how they can best capture it. McKinsey & Company (2016)
13. Asset management reimagined. Mainstream research report (2019)
14. Landeta, J.: Current validity of the Delphi method in social sciences. Technol. Forecast. Soc. Chang. **73**, 467–482 (2006)
15. Okoli, C., Pawlowski, S.D.: The Delphi method as a research tool: an example, design considerations and applications. Inf. Manage. **42**, 15–29 (2004)
16. Stentoft, J., Adsbøll Wickstrøm, K., Philipsen, K., Haug, A.: Drivers and barriers for industry 4.0 readiness and practice: empirical evidence from small and medium-sized manufacturers. Prod. Plann. Control **32**, 811–828 (2021)

17. Cichosz, M., Wallenburg, C.M., Knemeyer, A.M.: Digital transformation at logistics service providers: barriers, success factors and leading practices. Int. J. Logist. Manage. **31**, 209–238 (2020)
18. Schroeder, A., Ziaee Bigdeli, A., Galera Zarco, C., Baines, T.: Capturing the benefits of industry 4.0: a business network perspective. Prod. Plann. Control **30**, 1305–1321 (2019)
19. Peillon, S., Dubruc, N.: Barriers to digital servitization in French manufacturing SMEs. Procedia CIRP **83**, 146–150 (2019)
20. Trappey, A.J.C., Trappey, C.V., Hareesh Govindarajan, U., Chuang, A.C., Sun, J.J.: A review of essential standards and patent landscapes for the Internet of Things: a key enabler for industry 4.0. Adv. Eng. Inf. **33**, 208–229 (2017)
21. Gajic, S., Palcic, I.: A new framework for complexity analysis in international development projects – results from a Delphi study. Adv. Prod. Eng. Manage. **14**, 225–238 (2019)
22. Maletič, D., Almeida, N.M., Gomiščček, B., Maletič, M.: Understanding motives for and barriers to implementing asset management system: an empirical study for engineered physical assets. Prod. Plann. Control (2022, Accepted for publication)
23. Gavrikova, E., Volkova, I., Burda, Y.: Strategic aspects of asset management: an overview of current research. Sustainability **12**, 5955 (2020)
24. Maletič, D., Maletič, M., Al-Najjar, B., Gomiščček, B.: Development of a model linking physical asset management to sustainability performance: an empirical research. Sustainability **10**, 4759 (2018)
25. Alsyouf, I., Alsuwaidi, M., Hamdan, S., Shamsuzzaman, M.: Impact of ISO 55000 on organisational performance: evidence from certified UAE firms. Total Qual. Manage. Bus. Excell. **32**, 134–152 (2021)
26. ISO 55001 Asset management—Management systems—Requirements. International Organization for Standardization (ISO), Geneva, Switzerland (2014)
27. von Leipzig, T., et al.: Initialising customer-orientated digital transformation in enterprises. Procedia Manufact. **8**, 517–524 (2017)

Household Appliance Lifecycle and Sustainable Development

Adam Hamrol, Marta Pawelczak, and Filip Osiński(✉)

The Faculty of Mechanical Engineering, Poznan University of Technology, Poznań, Poland
filip.osinski@put.poznan.pl

Abstract. Economic growth is stimulated by a continuous increase in goods consumption. With regard to household appliances it manifests itself in its frequent replacement. This may have an impact on negative environmental and climatic changes. In this article, the authors deal with this problem in the context of consumer awareness of the impact of their behavior on these changes. The aim of the article is to check whether frequent replacement of household appliances is justified from the point of view of environmental pollution and climate protection. The article highlights not only the carbon footprint of two types of devices, but also consumer preferences for replacing old devices with new ones. As part of the research, a questionnaire was conducted on the frequency of replacement of household appliances and the reasons for their replacement. An analysis and comparison of the carbon footprint of two refrigerators - the old type and in line with the latest energy consumption standards - was carried out. It was found that the replacement of the refrigerator, due to the lower electricity consumption in the use phase, is ecologically profitable compared to the old type refrigerator after 8 years of use.

Keywords: Carbon footprint · Life Cycle Assessment · Environmental impact

1 Introduction

Household appliances were used in the past longer than today. The main reason was the high purchase price. Today, along with the increase in availability, this equipment has become relatively cheap, and consumers replace it relatively often – often more than is justified. It is commonly believed that this poses a significant threat to the implementation of a sustainable development strategy [1, 2].

The negative impact of household goods on the natural environment begins at the production stage, where many factors potentially threaten the natural environment, including:

- consumption of raw materials (metals, paints, glues, etc.)
- media consumption (water, energy)
- fossil fuel consumption (natural gas, heating oil)
- emission of pollutants into water and air
- waste generation

A. Hamrol et al. (Eds.): MANUFACTURING 2022, LNME, pp. 203–213, 2022.
https://doi.org/10.1007/978-3-031-00218-2_17

The mentioned environmental impacts are limited by a number of legal provisions. Production plants are required to update their permits in the event of launching a new installation, introducing new packaging or changing the preparations used.

The use of household appliances also has a negative impact on the natural environment, mainly due to energy consumption. The consumer is informed about this aspect of using the equipment via the energy label. One of the goals of its introduction was to influence consumer natural environment-oriented decisions about the choice of equipment.

The first energy labels appeared on cooling equipment in the 90s, and the E energy efficiency class was the dominant one. As a result of technological progress, the equipment achieved better and better classes over the next years. When the standard scale turned out to be insufficient, a plus sign was added to the classes. In this way, well-known classes such as A+, A++, A+++ were created. In the second decade of the 21st century, the classification method presented above became inadequate. In order to create space for further technological development, the European Commission decided to upscale energy efficiency and return to the original A-G scale [3]. The introduced change is to ensure better legibility of the labels. The new formula became valid on March 1, 2021 for three types of appliances: refrigeration equipment, dishwashing machines and washing machines. The most energy-efficient devices remain in the "green" class, and those with lower energy efficiency in "orange" and "red" classes. In practice, the green class is empty today and is waiting for new technological challenges of energy-saving and environmentally friendly solutions.

The lifetime of household appliances varies greatly. It depends to a large extent on the individual attitudes of consumers. However, the policy of producers also has a significant influence on lifetime [4]. In general, there is a noticeable tendency to shorten the life of household appliances. Technological development and fashion in a given industry force the displacement of old products in favor of new ones. This is caused by their wear or the growing expectations as to the effectiveness of the product. This is directly related to the marketing definition of a product life cycle, which says that a product is offered on the market as long as there is a demand for it [5].

Enterprises more and more commonly pursue a policy of purposefully shortening the life of household appliances, known as planned obsolescence (planned aging of products) [6]. It consists in introducing new, more technologically advanced equipment to the market, which at the same time displaces older equipment from the market (not necessarily worse) [7–9]. For this purpose, various marketing campaigns are used to motivate consumers to buy new products. In the case of electronic equipment, this is also achieved by programming that causes the equipment to stop working after a certain period of time or after a certain number of cycles. Such situations usually take place after the warranty has expired. In dishwashers or ovens, a way to deteriorate the operation of equipment after a certain period of time is to use components which, as a result of regular exposure to high temperature, wear out with a certain intensity [10].

The introduction of new regulations, which force changes and causes the removal of older equipment from the market, also works in a similar way to the planned obsolescence. Consumers voluntarily replace older, functional equipment with newer equipment. In addition to the requirements of the European Union regulations, customer requirements remain an important issue. The growing awareness of consumers has caused an increase in both the quality requirements and the requirements for the production and disposal of the purchased products [11, 12].

After the operation stage, household appliances in most cases are handed over by consumers to electronic waste collection points, which pass them for disposal or recycling. The basic stage of recycling is manual disassembly of each product. This allows for the removal of hazardous components and the segregation of reusable parts. The procedures of disassembly vary depending on the type of equipment as well as the number of hazardous components. The materials used in the product are recovered depending on the ease, cost and possibility of later development. In many cases, metals are collected first, then plastics, and finally composite materials and laminates. Materials are sorted according to shape, size and density.

The law imposes other obligations on producers related to environmental protection, consisting in achieving an appropriate level of recycling and recovery of:

- used electric devices and electronics
- packaging waste
- used batteries
- used oils

A separate issue is the repair of broken or damaged equipment. Theoretically, it should benefit consumers, first of all, and the natural environment through lower waste generation and better use of resources.

The research conducted in Rosko 2018 shows that 64% of respondents always repair equipment that has failed. The main reason why consumers did not repair the equipment turned out to be the high price of the repair. The other reasons are the preference to obtain new equipment and the feeling that the old equipment was outdated or obsolete [13]. The research identified the following elements that limit the availability of repair:

- lack of access to spare parts, technical information, diagnostic software and training, especially for independent repair shops; failure to ensure the availability of spare parts throughout the life of the equipment by the manufacturer
- no standardization of key components between brands (in particular, this applies to household appliances) or between appliances of the same brand,
- lack of technical knowledge necessary to carry out the repair due to the increasing complexity of equipment and the increasing share of electronic components and their miniaturization
- equipment design that prevents repair (e.g. disassembly of glued elements, welded plastic elements in washing machines, inaccessible screws, or non-standard screws)
- unattractive repair price due to the high cost of labor - often producing new equipment based on mass production is cheaper for the manufacturer

The research also highlighted the low profitability of hardware repair companies, which makes it more difficult for consumers to access such services.

2 Objectives of the Work and Research Method

The considerations presented in the introduction allow for the formulation of the following conclusions:

- consumers are replacing household appliances more and more often because they want to, because they can afford it, and because they are forced to do so by the manufacturer's policy of deliberately reducing the durability of the equipment
- it is not profitable for consumers to repair damaged equipment because it is either technically impossible or unprofitable
- equipment is becoming more energy efficient and more environmentally friendly

Therefore, the question arises whether the observed shortening of the service life of household appliances by manufacturers is justified economically and environmentally. The research described in this article aims to answer at least a part of this question.

The research was carried out in two stages. In the first stage, questionnaire research was carried out, the purpose of which was to obtain answers to the following questions:

- After what period of use do consumers replace household appliances?
- What are the reasons for replacing the equipment?

A special and structured research questionnaire was prepared. The questions were limited to the so-called large household appliances (e.g. refrigeration equipment, cooking equipment, washing and drying equipment).

A seven-point scale was used in the lifetime responses: from less than 1 year to more than 15 years. The questions relating to the reasons for the exchanges contained seven possible options to choose from, selected by the authors on the basis of their experience:

- malfunction
- home renovation
- high energy consumption
- lack of functions
- furnishing a new home
- the equipment is outdated
- aesthetic considerations

The respondents could select any number of responses. The survey was carried out using Google Forms in the period of May 1–August 8, 2021. A group of 465 respondents took part in the survey.

A refrigerator was selected as a representative of household appliances for the second stage of the research. Its environmental impact throughout its life cycle has been determined for production and use stage. Two types of refrigerators differing in technological advancement and energy consumption were considered.

The most common idea to measure the environmental impact is product carbon footprint (PCF). Product Carbon Footprint is the amount of carbon dioxide or its equivalent released to the atmosphere during all Life Cycle Assessment of the given product [14].

Because emissions during the production process are not only CO_2, other gases have been converted into CO_2eq with the indicator of global warming potential (GWP) in comparison to carbon dioxide [15]. The use of such a conversion factor allows for the unification of the results, regardless of the type of emission that occurs in a given process. The use of one unit in the form of CO_2Eq allows you to compare with each other not only similar products, as in the example below, but also products or processes significantly different from each other.

The calculations were based on the data available in the literature on GHG emission, energy consumption as well as ready CO_2eq emission indicators for raw materials, production processes and energy use [16, 17]. In all processes, indexes and values are based on the European Union economy, which is adequate for the production processes carried out in the European Economic Community (EEC). The calculations refer to the energy consumption for the production of individual components described in Table 1, but also to the total energy consumption in the production plant where the refrigerators described in the example were produced. This allows for a relatively accurate description of the impact of individual pieces of product, taking into account the maintenance of machinery and equipment, buildings or design, and planning work in the factory. For manufacturing processes, the CO_2eq has been adopted at the level of 0.836 – the factor calculated for Poland in 2017 [18].

In the last stage, it was analyzed how long the purchase of a new refrigerator with high energy efficiency pays off from the point of view of the consumer and the natural environment. This allowed assessment of whether the period after which consumers most often replace refrigerators is justified from the presented points of view.

3 Results of Consumer Surveys

In response to the question about the lifetime of household appliances, the answer that received the most votes was the range "average (5–10 years)", followed by the answer "quite rarely (10–15 years)". Microwave ovens were an exception, where the second answer that received the most votes from the respondents was "quite often (1–5 years)". The frequency and reasons for replacing various household appliances are described in Fig. 1.

In response to the question about the reasons for replacement of equipment, the most common answer was "failure". The second most common answer was the renovation of the kitchen/apartment/house, and the third most common answer was high energy consumption. The answer with the fewest indications was aesthetic considerations. The reasons for the replacement of household appliances are described in Fig. 2.

The economic analysis of potential customers was not taken into account because the amount of income of a given household does not translate into the impact of the equipment used on the natural environment and climate. The subject of the research was the direct impact of the household appliances used on the environment. The analysis of ecological behavior of consumers in relation to their earnings may constitute a separate topic.

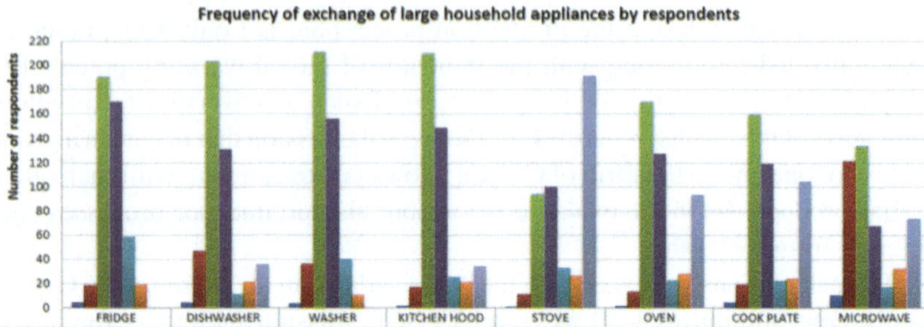

	FRIDGE	DISHWASHER	WASHER	KITCHEN HOOD	STOVE	OVEN	COOK PLATE	MICROWAVE
Often (< 1 year)	5	5	4	2	1	2	5	11
Quite often (1 - 5 years)	19	47	37	18	12	14	20	122
Average (5 - 10 years)	191	204	212	210	95	170	160	135
Quite rarely (10 - 15 years)	170	132	156	149	101	128	120	68
Rarely (> 15 years)	59	12	41	26	33	23	23	18
I do not change	20	22	12	22	27	29	25	33
Not applicable	0	36	0	35	192	94	105	74

Fig. 1. Frequency of large household appliance replacement.

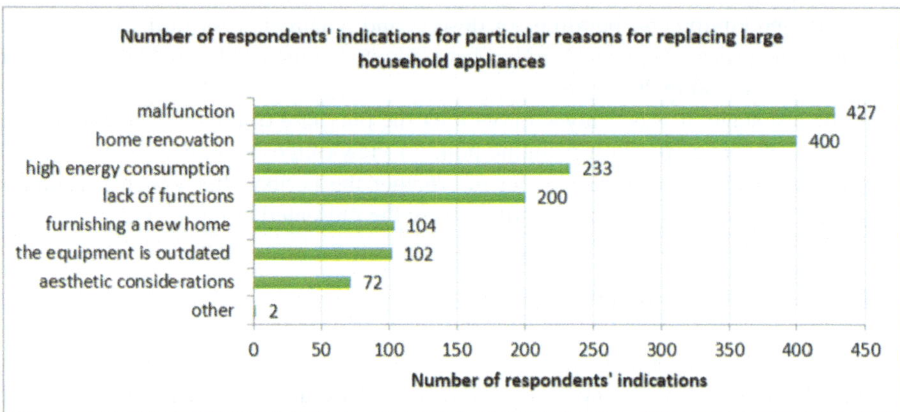

Fig. 2. Reasons for replacing large household appliances.

4 Carbon Footprint Analysis

The survey results suggest that it is advisable to ask the question: after how many years it pays off to replace old equipment with new one, taking into account that the new equipment is more energy-efficient than the old one, but its production causes an additional burden on the environment.

To reply to this question, an analysis of the carbon footprint was used.

A refrigerator was selected for comparative analysis because it consumes the most energy of other equipment due to working continuously. Two types of refrigerators were selected based on the European Union energy consumptions scheme – one with the lower energy efficiency class "F" and the second with the higher class "C".

The following assumptions were made:

- two built-in refrigerators (for domestic use) in energy classes “F” and “C”
- a trouble-free life cycle of both refrigerators - no part replacement, repair and refrigerant escape
- similar CO_2 emissions related to the transport of equipment - will be wrinkled in comparative analysis
- similar CO_2 emissions related to recycling - due to a similar construction of both equipment, the values obtained would be close - will be omitted in the comparative analysis

Emissions during production are the sum of the partial carbon footprints for semi-finished products and the refrigerator parts, and the carbon footprint of the energy use in the factory that produces the refrigerator. Partial carbon footprints are marks for a specific component made of a known material. The estimated weight of a component was multiplied by the value of the benchmark, thus obtaining its carbon footprint. All partial carbon footprints were summed [19–21].

The amount of CO_2 emissions directly related to production processes was calculated on the basis of the total amount of energy used in the factory and the production volume. This approach allowed to take into account all direct and indirect sources of energy consumption. Thanks to this, the carbon footprint of the presented products includes elements such as maintenance of machinery, internal transport or maintenance of buildings in which the production process takes place.

The unit energy consumption per 1 refrigerator was determined, and then, knowing the amount of energy consumed in the plant, the total amount of CO_2 emitted was estimated. Annual emission of CO_2 equivalents was calculated by multiplying CO_2 emission per kWh for Poland and the electricity consumption for both refrigerators. The service life was set at 10 years and multiplied by the indirect annual emissions resulted in the total lifetime CO_2 emissions of both refrigerators.

The CO_2 emissions during production for both refrigerators - in the higher and lower energy efficiency classes - are similar. On the other hand, when comparing the CO_2 emissions resulting from use, the refrigerator in the higher energy efficiency class, class “C”, was almost twice as low, which in turn translated into twice lower emissions throughout its life cycle.

Taking into account the environmental impact of both the production and use of a new refrigerator in energy efficiency class “C”, compared to the old device in class “F”, it can be recognized when the given values balance. If you consider the option of using an existing refrigerator, the indirect emission of which is 240 kg CO_2eq, and the purchase of a new one, the production of which will emit 917 kg CO_2, and the annual use of 123.7 kg CO_2eq, the replacement pays off after 8 years of use. If it is necessary to buy a new refrigerator, it is generally advisable to buy a device in the best energy efficiency class possible, because CO_2 emissions in the production process are largely similar for all types of devices. The replacement balance of an already owned refrigerator is shown in Fig. 3.

Table 1. Fridge production and use carbon footprint [19–21]

		Fridge in a lower energy efficiency class				Fridge in a higher energy efficiency class			
	Volume	267 L				264 L			
	Energy efficiency class	F				C			
	Dimensions [cm]	175.5 × 54 × 55				175.5 × 54 × 55			
	Element	Raw material	Weight [kg]	Emission index CO_2 [CO_2/kg]	Partial carbon footprint [CO_2/kg]	Raw material	Weight [kg]	Emission index CO_2 [CO_2/kg]	Partial carbon footprint [CO_2/kg]
Raw material	Casing	Stainless steel	14.5	6.15	89.2	Stainless steel	14.5	6.15	89.2
	Assembly reinforcements	Carbon steel	1.0	1.77	1.8	Carbon steel	1.0	1.77	1.8
	Hinges	Stainless steel	1.2	6.15	7.4	Stainless steel	1.2	6.15	7.4
	Insulation	PUR 5501	5.5	4.84	26.6	HDPE vacuum panel	3.0	1.10	3.3
						PUR 1801	1.8	4.84	8.7
		Aluminum foil	0.5	11.89	5.9	Aluminum foil	0.5	11.89	5.9
	Evaporator	Aluminum	2.0	11.89	23.8	Aluminum	2.0	11.89	23.8
	Condenser	HDPE	1.0	1.10	1.1	HDPE	1.0	1.10	1.1
	Hot and cold pipe	Cooper	2.0	2.77	5.5	Cooper	2.0	2.77	5.5
	"No Frost" canal	PPE	2.0	2.40	4.8	PPE	2.0	2.40	4.8
	Compressor	Carbon steel	2.3	1.77	4.1		2.7	1.77	4.8
		Aluminum	1.5	11.90	17.9	Aluminum	1.5	11.90	17.9
		Mineral oil	0.5	1.07	0.5	Mineral oil	0.5	1.07	0.5
	Base for the compressor	Galvanized steel	0.5	1.99	1.0	Galvanized steel	0.5	1.99	1.0
	Refrigerant	R290 (propane)	0.7	2.86	2.0	R290 (propane)	0.7	2.86	2.0
	Inside	HDPE	6.5	1.10	7.2	HDPE	6.5	1.10	7.2
	Door seal	PVC	1.0	2.22	2.2	PVC	1.0	2.22	2.2
	Balconies and drawers in the refrigerator	PPE	2.7	2.40	6.5	PPE	2.4	2.40	5.8
	Drawers in the freezer	PP	1.3	1.95	2.5	PP	1.3	1.95	2,5
	Shelfs	Tempered glass	5.0	0.85	4.3	Tempered glass	5.0	0.85	4.3

(continued)

Table 1. (*continued*)

		Fridge in a lower energy efficiency class				Fridge in a higher energy efficiency class			
SUM			51.7		214.2		51.1		199.6
Carbon footprint of a single item		740.5				720			
Emission during production [kg CO_2eq]		953.0				917.9			
Usage phase	Electricity consumption [kWh/year]	288				148			
	Kg CO_2eq/kWh	0.836 [28]							
	Annual indirect emission [kg CO_2eq/year]	240.8				123.7			
	Lifetime	10 years				10 years			
Total CO_2 Eq [kg CO_2eq]		2407.7				1237.3			

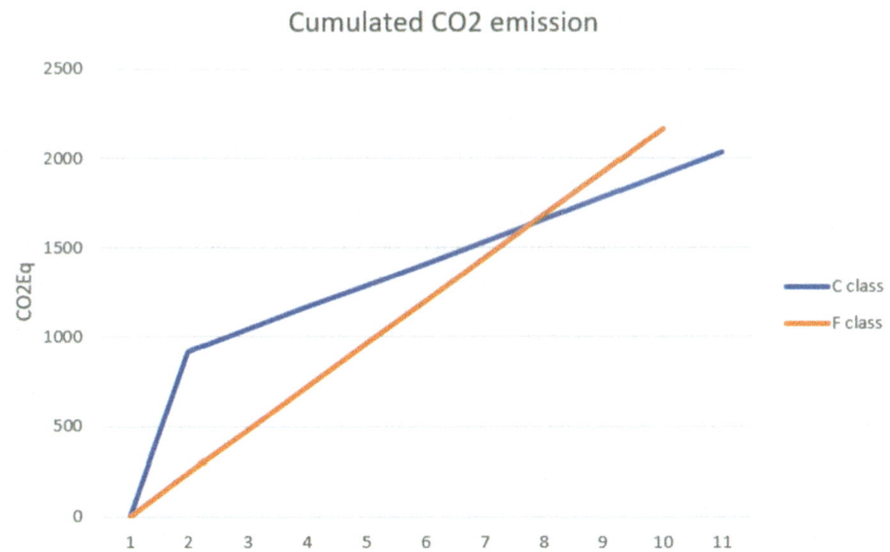

Fig. 3. Comparison of a new refrigerator in energy efficiency class "C" with the use of an existing one in class "F".

5 Conclusions

The main reason for replacing a household appliance is its failure. But energy consumption also ranks high. Using the example of a refrigerator, it has been shown that after 8 years of use, repairing the refrigerator is not profitable, both for environmental and economic reasons. Most of the respondents replace the equipment at a time when its replacement is justified due to the CO_2 balance.

The production of a refrigerator in energy efficiency class "C" is more ecological, mainly due to the lower use of PUR foam, which is a harmful factor, and is visible in the assessment of the carbon footprint for the production process (historically, the carbon footprint was significantly influenced by the abandonment of HCF in the 1990s). In addition, the four years of operation of a refrigerator in energy efficiency class "F" offsets the carbon footprint needed to produce a refrigerator in class "C" – after this period, we start to "save" the natural environment.

If the consumer has a working refrigerator, the difference in energy consumption will make it ecologically viable to replace it after 8 years of use. The relatively long time necessary to balance the environmental impact of replacing a given equipment should aim to shorten it as much as possible. We can get it in three ways:

- by reducing the product carbon footprint - e.g. by reducing the use of natural resources, with particular emphasis on substances that have a negative impact on the climate - e.g. the ban on the use of freon
- by extending the life cycle of a given product - e.g. ensuring the availability of spare parts and enabling repairs in the event of minor failures
- by reducing the energy consumption of the finished product - which is in fact constantly happening for many years due to customer requirements

In many cases, the rapid development of technology and the implementation of new products for sale make it difficult to clearly determine whether the replacement of specific equipment is justified from the point of view of climate protection. A strong share of marketing in the activities of enterprises encourages consumers to make purchases, which are not always environmentally justified, but only constitute greenwashing. The conducted research may constitute an introduction to the further development of the topic: the impact of product development and the implementation of new production standards on the natural environment.

References

1. Household appliances sector. https://www.paih.gov.pl/sektory/agd. Accessed 28 Mar 2021
2. Household appliances industry. https://applia.pl/agd-info/polska/. Accessed 28 Mar 2021
3. WEEE Label. https://europa.eu/youreurope/business/product-requirements/labels-markings/weee-label/index_pl.htm. Accessed 27 Apr 2021
4. Lifetime. https://wsjp.pl/index.php?id_hasla=15413&id_znaczenia=5066753&l=9. Access 27 June 2021
5. Koper, K.: The structure of the life cycle management model of technical objects. Ph.D dissertation, Poznan University of Technology, Poznan, pp. 17–19 (2016)
6. Ryś, A.: Planned aging of the product - analysis of the phenomenon. ZN WSH Management, pp. 121–128 (2015)
7. European Comission: Sustainable de–growth: an alternative to sustainable development? DG Enviroment News Alert Service, p. 1 (2010)
8. Rokicka, E., Woźniak, W.: Towards sustainable development. Concepts, interpretations, contexts. Katedra Socjologii Ogólnej, Łódź (2016)

9. Stawicka, E.: Corporate social responsibility and sustainable production and consumption. Assoc. Agric. Agribus. Econ. Sci. Ann. **XVI**(3), 265–269 (2014)
10. Kłos, L.: Bulb conspiracy - the contemporary dimension of planned unsuitability, Studies and Works WNEIZ US, nr 53/1, pp. 8–9 (2018)
11. Hamrol, A.: Quality Management with Examples. Scientific Publisher PWN, Wyd. II, Warsaw, pp. 19–20 (2007)
12. Marszałek, A.: Shaping quality in the product life cycle. In: Salerno-Kochana, M. (ed.) Selected Aspects of Quality Management, pp. 167–168. Cracow University of Economics, Kraków (2017)
13. European Parliamentary Research Service: Consumers and repair of products (2019)
14. ISO 14067:2018 Greenhouse gases—carbon footprint of products—requirements and guidelines for quantification and communication (2018)
15. AR5 Synthesis Report: Intergovernmental Panel on Climate Change (2015)
16. Cushman-Roisin, B., Tanaka Cremonini, B.: Useful numbers for environmental studies and meaningful comparisons useful numbers for environmental studies and meaningful comparisons. Dartmouth College (2019)
17. Łasut, P., Kluczycka, J.: Methods and programs for calculating the carbon footprint. Institute of Mineral and Energy Economy of the Polish Academy of Sciences, nr. 87 (2014)
18. Country specific electricity factors, p. 3 (2018). https://www.carbonfootprint.com/docs/2018_8_electricity_factors_august_2018_-_online_sources.pdf. Accessed 30 Sept 2021
19. https://www.carbonfootprint.com/docs/2018_8_electricity_factors_august_2018_-_online_sources.pdf
20. Emission factors in kg CO_2-equivalent per unit. https://www.winnipeg.ca/finance/findata/matmgt/documents/2012/682-2012/682-2012_Appendix_H-WSTP_South_End_Plant_Process_Selection_Report/Appendix%207.pdf. Accessed 30 Sept 2021
21. Carbon footprint of the hot-dip galvanisation process using a life cycle assessment approach. https://www.sciencedirect.com/science/article/pii/S266679082100001X. Accessed 30 Sept 2021

Waste Minimization in the Battery Assembly Process - Case Study

Agnieszka Kujawińska(✉), Adam Hamrol, and Krzysztof Brzozowski

Poznan University of Technology, Piotrowo 3 Street, 61255 Poznan, Poland
agnieszka.kujawinska@put.poznan.pl

Abstract. The article presents a case study of the improvement of the car battery assembly process. The aim of the improvement activities was to minimize waste, which had a significant impact on the natural environment. The DMAIC (Six Sigma) approach was used in the improvement actions. The solutions implemented in the process brought measurable benefits for the company.

Keywords: Six Sigma · DoE · Assembly process

1 Introduction

One of the many pillars of the world around us is energy. Modern man cannot imagine civilization without utilities such as electricity, hot running water or heating. It is an "existential standard" whose even a temporary lack causes discontent, stress, and sometimes even rebellion assuming the form of a strike. People are used to coexisting with utilities which they treat as part of their everyday life. For decades, people have taken little heed of energy sources that are a boon to civilization [1, 2]. It was only at the beginning of the 21st century that the problem of their non-renewal was addressed. The depletion of natural resources is currently the subject not only of discussions, but also measures which are supposed to actively influence the saving of raw materials and reduce the negative impact of all processes on the surrounding environment.

Sustainable development assumes the interdependence and equivalence of three areas: economy, nature and society, which makes it possible to search for ways to deal with contemporary threats that result from the destruction of environmental resources or social shocks [3, 4]. Changing the relationship between these areas for the better requires, among others, the use of more environmentally friendly technologies and the elimination of forms of production which have a negative impact on them. Another factor behind positive changes may be the reduction of consumption and a change in the current value system oriented towards possessing material goods. In the area of economics and finance, the biggest barrier that hinders the introduction of the sustainable development trend in many countries is the dominance of a development model based on economic growth, which is seen as a priority. Human rights, well-being and environmental constraints are no less important.

In developing countries, there is a common view that investing in environmental protection should be implemented at a later stage of development, which in practice

A. Hamrol et al. (Eds.): MANUFACTURING 2022, LNME, pp. 214–224, 2022.
https://doi.org/10.1007/978-3-031-00218-2_18

means accepting environmental degradation to meet current needs. In order to counteract that, it is necessary to integrate economic, environmental and social aspects. The concept of sustainable development is based on market strategies that are based on technologies and legal regulations. In essence, it is to direct the world towards sustainable development without many radical changes that would destroy the current social order. In addition, this idea assumes a thesis that the currently used production technologies are the source of environmental threats [5–7].

A particular attention is paid to manufacturing processes, which largely contribute to the exploitation of deposits and often have a negative impact on the environment. On the one hand, one strives after designing products that consume less energy and require less raw material to make them. On the other hand, one makes an effort to ensure that production processes themselves are not a source of waste and that their energy consumption is reduced [8, 9]. It is of particular importance in processes where waste is harmful to the environment. An example of such a process is the manufacturing process of automotive batteries.

2 Research Object: Battery Manufacturing Process

In terms of construction, a battery is a set of cells connected electrically and closed in separate chambers inside a casing. Batteries generate voltage as a result of chemical reactions that take place in them. Depending on the type of battery, various elements are found in them, including those that are harmful and dangerous to the environment and human health and life.

In the process of producing batteries, a number of substances are used that affect safety and functionality, which may be harmful to the environment. The most commonly used are: manganese dioxide, iron, zinc, graphite, ammonium chloride, copper, potassium hydroxide, mercury, nickel, lithium, cadmium, silver, cobalt, as well as glass, silica, paper, foil and hydrogen. Heavy metals found in batteries (lead, cadmium and mercury) have a negative effect on health, and acids or bases used to form electrolyte are caustic and corrosive. Therefore, producers observe strict environmental rules relating to the manufacturing process. The same happens during one of the last stages of the process - battery assembly.

Depending on the kind and type of battery, the assembly process consists of a number of sub-processes (Fig. 1).

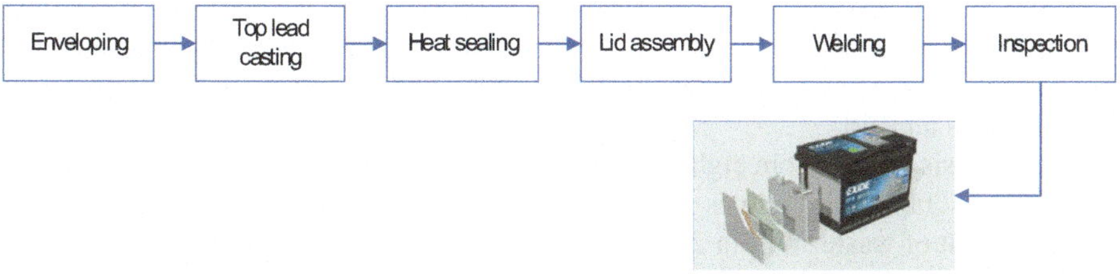

Fig. 1. Flowchart of the battery assembly process (own study).

The first is the so-called enveloping. It consists in the production of sets composed of groups of lead plates of opposite polarity (negative and positive current collectors) arranged alternately and isolated from each other by means of a microporous separator. The separator isolates physically the positive and negative plates (protects against short-circuits) and allows the flow of electrons during battery operation.

The next stage is the so-called top lead casting. The role of this sub-process is to connect separately a series of lugs of positive and negative plates, which will constitute one of the battery cells. The operation consists in casting the bridges by immersing the lugs of the battery sets in a mold (sockets) filled with lead, cooling the casting and stamping it out. 6 sets of plates prepared in this way will be placed in separate compartments of the battery block.

The next sub-process is heat sealing between the bulkheads of the battery cells to produce successive bridge connections. The head of the welding torch, armed with point electrodes, connects the flaps of the bridges, generating energy that enables a local weld. The next stage of assembly consists in tightly connecting the battery block to the lid. A hot plate with an appropriate shape fuses the edges of the block and the lid. Such a heated copolymer is then joined under appropriate pressure to form a tight seal.

One of the last major subprocesses is the welding of bushings. It aims to create a tight and mechanically durable connection of the pin - extension the battery set with the lid sleeve. The remaining operations are related to quality inspection, which include, among others, leakage testers, battery height and pole height testers.

The currently used technology minimizes harmful impact on the natural environment. Unfortunately, the emerging problems related to the risk of a negative impact on the environment are related to the instability of the process. The effect of instability is the risk of waste formation. This means wasted energy, material, working hours and other resources. A significant part of the loss is related to material (lead, cadmium, plastic, etc.). In order to minimize losses, improvement projects are implemented to increase the stability and predictability of processes.

3 Case Study: Problem of Excessive Waste in the Battery Assembly Process

The battery assembly stage is one of the key steps in the entire production process. The purpose of the activities carried out at this stage is to obtain a battery compliant with the requirements which is transferred to the final stage - filling with electrolyte. Therefore, it is crucial to maintain all the required properties of the product. One of them is the tightness of the assembled battery. Meeting the requirements related to tightness is important from the point of view of product safety and its impact on the environment - electrolyte is an acid.

The analysis of data from eight assembly lines revealed the problem of an unacceptable level of scrap on three lines - over 0.33%. Lines 7, 6 and 1 were characterized by a large amount of waste. Within 2 months, these lines generated approximately 92% of scrap in the entire assembly process (Fig. 2). The main nonconformities which caused the battery to be considered defective was the lack of tightness.

The lack of tightness in a product means that it cannot be transferred to the next production stages. The product at this stage becomes a scrap which generates excessive production costs estimated at approximately EUR 29,000 per year. It becomes important to establish the causes of large amounts of waste on these lines.

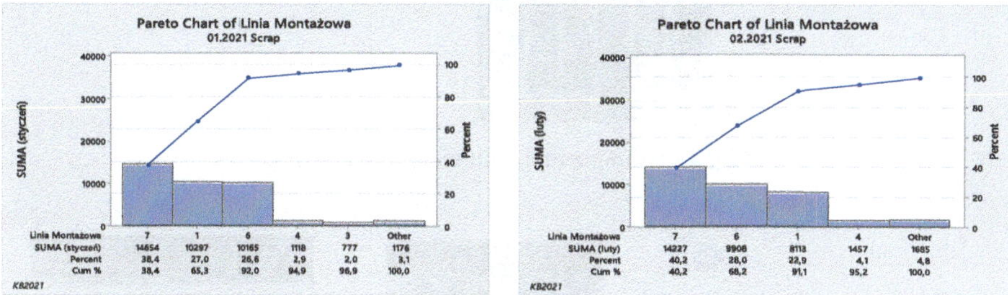

Fig. 2. Pareto chart for STD assembly (lines from 1 to 8) (own study).

In order to determine the potential causes of the product's leakage, a leakage test was performed for 47 batteries recognized as defective. The battery lid, which is designed to tightly seal the battery compartment, is divided into 56 areas. Leakage tests showed that 3 areas are prone to leakage: B5, E5 and F5 (Fig. 3).

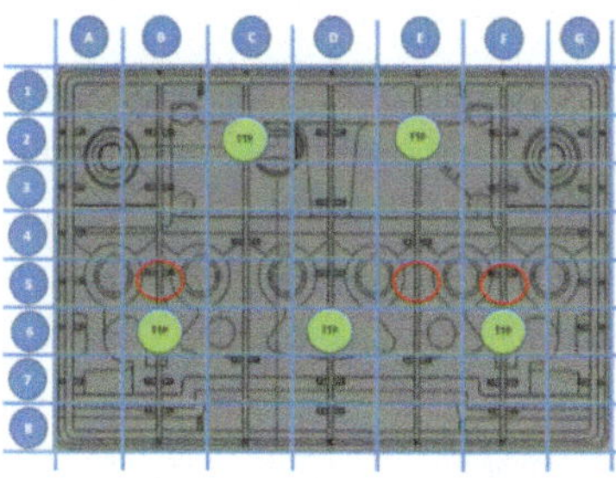

	A	B	C	D	E	F	G
1							
2							
3							
4							
5		32			4	13	
6							
7							
8							

Fig. 3. Division of the lid area in leakage tests (own study).

It turned out that most of the leaks were the so-called external leaks. It was found that internal leaks detected at the assembly stage did not represent a significant share of the total waste resulting from the process of welding the block to the lid. Leaks inside the battery resulted from a fault weld (Fig. 4) and constituted a small part of all nonconformities. It was decided for the project to focus on eliminating external leaks, which are mainly detected after the formation stage with a high voltage leak tester.

External leaks result from leaks in the lid put on the battery box. It was decided to verify the quality of their connection by indirect measurement method. The variable control was the force to be applied to the lid in order to break it. The measurement was performed on a testing machine, and the test sample consisted of elements of a welded block with a lid measuring 15 × 60 mm. In order to determine the magnitude of the breaking force in the present process, measurements were made in 4 areas located on each of the 4 side walls of the battery (Fig. 5).

Fig. 4. Example of a fault weld (own study).

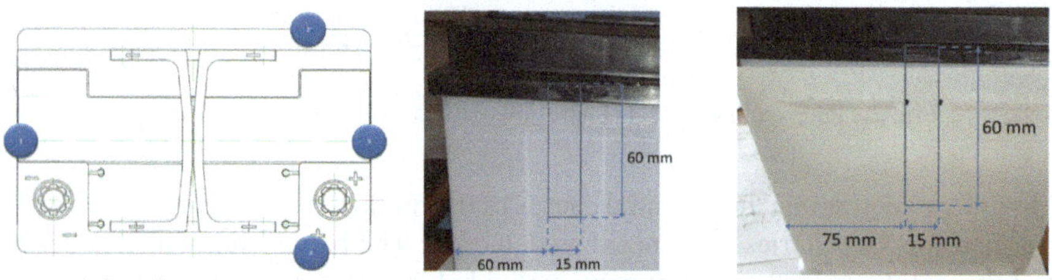

a) Sampling points b) Long side of the battery c) Short side of the battery

Fig. 5. Sampling points for testing (own study).

The analysis of the breaking force results allowed to indicate the magnitude of the force from which the joint will be considered tight. All samples with a strength greater than 630N will be considered as meeting the leakage condition (Fig. 6).

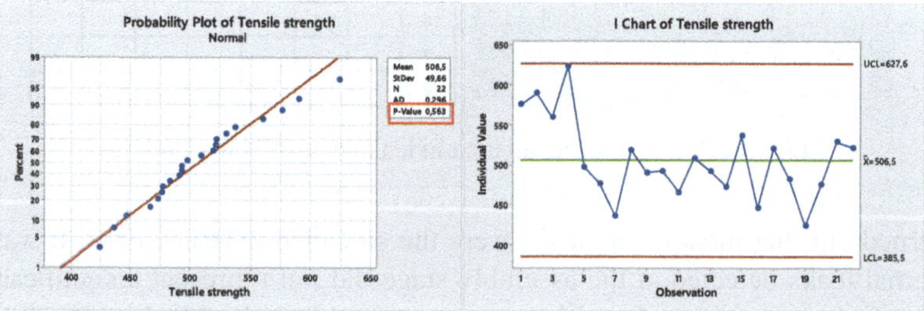

Fig. 6. Distribution of the breaking force measurement results (own study).

In order to identify the potential causes of leakage, an Ishikawa diagram was made on the basis of brainstorming (Fig. 7).

Then, an ABC analysis was performed for the selected causes. It turned out that 4 of them achieved the highest number of points. They were: deflection of the block wall by the impact of the so-called centering clamps, incorrect welding parameters, dirty melting plates, thickness of the plates.

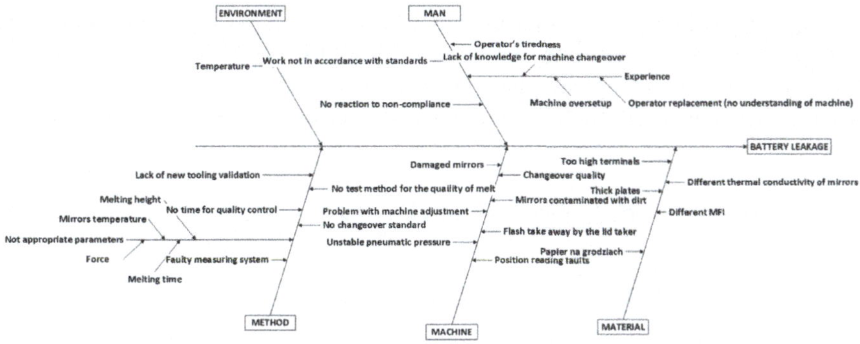

Fig. 7. Ishikawa diagram for a defect: leakage (own study).

In order to verify the hypothesis that the deflection of the wall of the battery block is caused by too much pressure of the clamps centering the block in the machine, a series of experiments was carried out in which it was examined whether the reduction of the clamp pressure would have a positive effect on the strength of the lid weld. A total of 16 samples were prepared: 8 were used with the current process parameters and 8 with new parameters (reduced pressure). Such a set of experiments was used to verify the null hypothesis H0: *All tensile strength means are equal*, with the alternative hypothesis: Ha: *Not all tensile strength means are equal.*

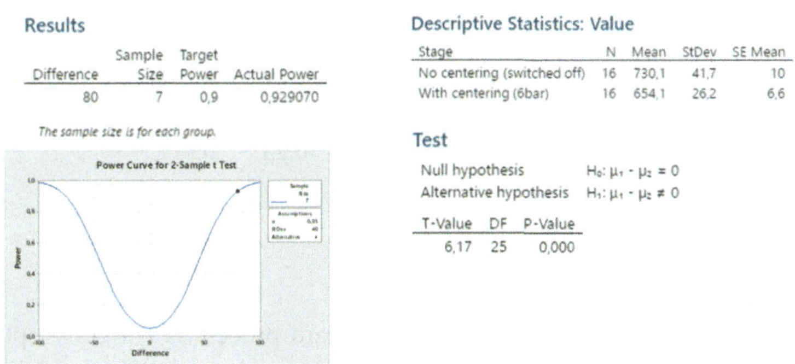

Fig. 8. Results of the data analysis from the experiment of the influence of the pressure force of the centering clamps (own study).

The analysis of the experimental results confirms that the force exerted on the centering clamps influences the strength of the connection between the lid and the wall (Fig. 8).

In order to verify another cause - welding parameters settings - an experiment was carried out to verify whether the currently used settings are appropriate to achieve the expected weld strength. For this purpose, three factors that had the greatest impact on the process of connecting battery components were selected. They were:

- melting point, which can be controlled by a set of heaters set into the heating plate.
- melting time, which is the time during which the block and lid material is heated in direct contact with the melting plate, the so-called mirror.
- penetration height, i.e. the depth of melting of a given component with a heated melting plate.

The experiment was performed for 3 levels of factors (Table 1).

Table 1. Level values for three selected factors (own study).

Factor	Levels	Values
Melting temperature [°C]	3	330, 355, 380
Melting time [s]	3	4, 5, 5, 5.5
Penetration height [mm]	3	1, 1, 5, 2

The experiment showed that one of the factors - penetration height, has a statistically significant influence on the quality of the block connection with the lid (Table 2).

Table 2. Results of experiment 3^3 (own study).

Term	Effect	T-Value	p-Value
Melting temperature [°C]	−94,7	−1,61	0,124
Melting time [s]	13,7	0,23	0,817
Penetration height [mm]	−159,5	−2,72	0,014

It was found that the value used in the standard process is correctly optimal and hence the machine operating parameters are adequate and have no effect on the increased battery waste.

In order to verify the influence of another potential cause on the quality of the weld, an element of autonomous maintenance was introduced, which consisted in standardizing and training operators in the method of cleaning melting plates. This activity was included in the shift duties of the line leader. Next, the proportion of waste before and after removing scales from the melting plates over equal time periods was compared (Table 3).

The results of the analyses showed that the dirt of the plates did not affect the strength of the connection between the block and the lid in critical places.

Table 3. Results of the comparison of the level of scrap before cleaning the plates and after introducing cleaning activities (own study).

Null hypothesis $H_0 : p_1 - p_2 = 0$
Alternative hypothesis $H_1 : p_1 - p_2 \neq 0$

Descriptive Statistics

Sample	N	Event	Sample p
Sample 1	91487	36	0,000393
Sample 2	241758	112	0,000463

Method	Z-Value	P-Value
Normal approximation	-0,89	0,376
Fisher's exact		0,461

In order to assess the influence of the thickness of the battery plates on the welding strength, the value of the breaking force for the lid welds in 24 batteries with a 0.9 mm (12 pieces) and 0.8 mm (12 pieces) separator was compared (Table 4).

Table 4. Results of the comparison of the breaking force of the lid for batteries with 0.9 mm and 0.8 mm thick plates (own study).

Descriptive Statistics

Sample	N	Mean	StDev	SE Mean
0,9 separator	12	689,8	43,1	12
0,8 separator	12	691,9	48,1	14

Null hypothesis $H_0: \mu_1 - \mu_2 = 0$
Alternative hypothesis $H_1: \mu_1 - \mu_2 \neq 0$

T-Value	DF	P-Value
-0,11	21	0,910

The results of the analysis show that there are no differences between the strength of the weld for 0.9 and 0.8 mm thick separators. Thus, it can be concluded that this is not the source of the battery leakage problems.

As a result of the preliminary analyses, it was decided to redesign the clamps centering the lid and the pressure on the centering clamps will be changed. For this purpose, an active experiment was performed, which allowed for the selection of an appropriate setting of the factors. Two factors were taken into account in the experiment: the type of clamps and the pressure of their tightness. Two types of clamps were used in the experiments, the so-called old and new (Fig. 9) and two pressure values: 2 Bar and 6 Bar.

The results of the analysis of the data from the experiment showed that the strongest connection was created with the use of new clamps and low pressure (Fig. 10, 11).

The evaluation of the pull-off force for batteries in which the lid was welded with the use of new clamps and a 2 Bar pressure also confirms the result of the experiment. The strength of this system is on average greater than the strength of the weld with the current method of joining.

It was decided that such a system would be put into production. The control of the effects of the implementation of the new method of connecting the lid to the battery consisted in observing the percentage of non-compliance on the analyzed assembly lines in 13 consecutive weeks from the introduction of changes. It turned out that the level of non-compliance decreased from 0.33% to 0.01%.

Fig. 9. Comparison of the old and new centering clamps (own study).

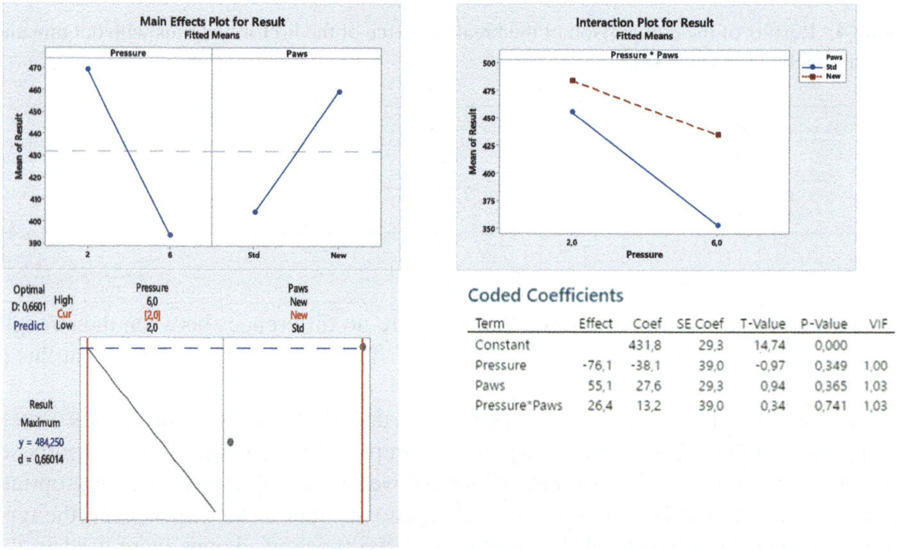

Term	Effect	Coef	SE Coef	T-Value	P-Value	VIF
Constant		431,8	29,3	14,74	0,000	
Pressure	-76,1	-38,1	39,0	-0,97	0,349	1,00
Paws	55,1	27,6	29,3	0,94	0,365	1,03
Pressure*Paws	26,4	13,2	39,0	0,34	0,741	1,03

Fig. 10. Visualization of the experiment results for two factors: type of clamps, pressure (own study).

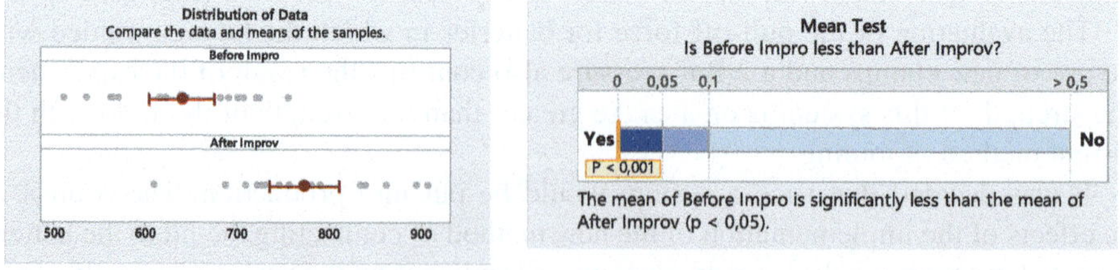

Fig. 11. Comparison of data distributions and the average strength of the weld for the current and new joining method (own study).

4 Summary

For most enterprises, maintaining customer satisfaction is a key factor in remaining competitive. Nowadays, when choosing a product, customers are more and more often guided not only by the price-related criterion, but also by the impact a product exerts on the environment, by information about the scale of the influence of company processes on the environment, and by the origin of resources and work safety.

The negative impact of manufacturing processes on the environment may result from the type of technology used and the instability of individual operations. The result of this instability may be the occurrence of losses in the form of a high level of non-compliant products (parts), which for the company is associated not only with a financial cost, but also with various types of threats related to e.g. employee safety, customer trust, environment, natural resources and many more.

Therefore, it is crucial to improve manufacturing processes by minimizing process losses. The presented example shows how a methodical approach to process improvement based on the instruments of quality management: statistical analysis and experiments planning contributes to the minimization of losses.

The proposed changes in the battery assembly process allowed for the minimization of material losses from 0.33% to 0.05%. The implemented changes also allowed to increase work safety and standardize control processes. The implementation of changes resulted in savings of around EUR 50,000 per year.

Acknowledgments. The paper is prepared and financed by scientific statutory research conducted by Chair of Management and Production Engineering, Faculty of Mechanical Engineering and Management, Poznan University of Technology, Poland, Poland, supported by the Polish Ministry of Science and Higher Education from the financial means in 2021 (0613/SBAD/4710).

References

1. Jasiulewicz-Kaczmarek, M., Drożyner, P.: Social dimension of sustainable development – safety and ergonomics in maintenance activities. In: Stephanidis, C., Antona, M. (eds.) UAHCI 2013. LNCS, vol. 8009, pp. 175–184. Springer, Heidelberg (2013). https://doi.org/10.1007/978-3-642-39188-0_19
2. Jasiulewicz-Kaczmarek, M., Saniuk, A.: How to make maintenance processes more efficient using lean tools? In: Goossens, R.H.M. (ed.) AHFE 2017. AISC, vol. 605, pp. 9–20. Springer, Cham (2018). https://doi.org/10.1007/978-3-319-60828-0_2
3. Starzyńska, B., et al.: Requirement's elicitation of passengers with reduced mobility for the design of high quality, accessible and inclusive public transport services. Manag. Prod. Eng. Rev. **6**(3), 70–76 (2015). https://doi.org/10.1515/mper-2015-0028
4. Kujawińska, A., Diering, M., Rogalewicz, M., Żywicki, K., Hetman, Ł.: Soft modelling-based methodology of raw material waste estimation. In: Burduk, A., Mazurkiewicz, D. (eds.) Intelligent Systems in Production Engineering and Maintenance – ISPEM 2017, vol. 637, pp. 407–417. Springer, Cham (2018). https://doi.org/10.1007/978-3-319-64465-3_39
5. Bożek, M., Kujawińska, A., Rogalewicz, M., Diering, M., Gościniak, P., Hamrol, A.: Improvement of catheter quality inspection process. In: 8th International Conference on Manufacturing Science and Education – MSE 2017 "Trends in New Industrial Revolution". MATEC Web Conference, vol. 121 (2017). https://doi.org/10.1051/matecconf/201712105002

6. Karwasz, A., Trojanowska, J.: Using CAD 3D system in ecodesign—case study. In: Golinska-Dawson, P., Kolinski, A. (eds.) Efficiency in Sustainable Supply Chain, pp. 137–160. Springer, Cham (2017). https://doi.org/10.1007/978-3-319-46451-0_9
7. Trojanowska, J., Karwasz, A., Machado, J.M., Varela, M.L.R.: Virtual reality based ecodesign. In: Golinska-Dawson, P., Kolinski, A. (eds.) Efficiency in Sustainable Supply Chain. ECOPROD, pp. 119–135. Springer, Cham (2017). https://doi.org/10.1007/978-3-319-46451-0_8
8. Starzyńska, B., Bryke, M., Diakun, J.: Human lean green method—a new approach toward auditing manufacturing & service companies. Sustainability **13**(19), 10789 (2021)
9. Baskiewicz, N., Barbu, C.: The role of internal conditions in the implementation of the lean green concept: American, Japanese and Polish experiences. Manag. Prod. Eng. Rev. **12**(4), 71–86 (2021)

Production Flow Between Workstations Using the Kanban and DBR Methods – Comparative Study

Klaudia Tomaszewska(✉)

Bialystok University of Technology, Wiejska 45A Street, 15-351 Bialystok, Poland
k.tomaszewska@pb.edu.pl

Abstract. Enterprises are constantly looking for methods and tools that will make them more efficient and competitive. The choice of good practice should be consistent with the individual goal of the enterprise. There is no one golden mean, as the dynamics and predictability of enterprises differ. It is very important to be consistent with the changes introduced, to keeping priorities and to be motivated to maintain them. The aim of the study is to present a comparison of two methods of organizing the production flow between workplaces, considering the management concepts under which methods were developed. This comparison includes the Kanban method (from Lean Manufacturing) used in the analyzed enterprise against using the Drum-Buffer-Rope method (from Theory of Constraints) instead of Kanban. The study presents, among other things, the similarities and differences between the methods in this comparative study. In addition, the scientific article contains proposals for further simulation studies of the application of the DBR versus Kanban method.

Keywords: Lean Manufacturing · Theory of Constraints · Kanban · Drum-Buffer-Rope · Case study

1 Introduction

Lean Manufacturing is one of the common, most frequently used method of production management in an enterprise. According to the definition, a Lean company builds its process management in such a way that the customer ordering a specific product only pays for its production, and not for the functioning of the enterprise [1]. Lean creates a work culture within the organization that makes all partners of the organization interested in striving for continuous improvement. Aimed at the elimination of waste, understood as: overproduction, waiting, inappropriate methods of production, unnecessary transport, unnecessary stocks, shortages, unused human potential. The main evil in system productions is waste [2]. The liquidation of losses increases the added value of activities carried out within the organization. Theory of Constraints (TOC) formulated in the 1970s by Eliyahu Goldratt, according to which an organization is as good as its weakest link. The most important concept of TOC is system constraints. Constraints can be physical in the form of resources, described as "bottle-necks", defined as a set of rules,

A. Hamrol et al. (Eds.): MANUFACTURING 2022, LNME, pp. 225–236, 2022.
https://doi.org/10.1007/978-3-031-00218-2_19

measures, or premises of the value of the basis on which the policy of the enterprise is developed. The bottleneck determines the theoretical production capacity, or the highest quality maximum production delivered to the market [3]. The bottleneck determines a system's output [4]. The idea behind TOC is concept that there is always one constraint, that hinders the throughput of any process. Depending on the flow, complexity of the production, that may be more than one resource limiting system performance (however, their number is always very small). Increasing the efficiency of the system will be possible when the weakest link is identified.

The aim of the article is to compare two methods of organizing the production flow between workplaces, the Kanban method, and the Drum-Buffer-Rope (DBR) method, considering the management concept under which these methods were developed. The most important, but not the only research question is the assessment of the effects when the Kanban system is replaced by the DBR method in a selected manufacturing enterprise.

The study contains introduces and presents theoretical issues related to Kanban and DBR methods. In addition, it contains information about the research tools used. The article uses a case study in an enterprise, and the purpose of the analysis was to identify bottlenecks in the process and to compare the application of the Kanban and DBR methods. Which was presented in the following chapters of this study, along with the similarities and differences between the analyzed methods, and conclusions.

2 Kanban Method

The Lean method uses several dozen complementary tools and techniques that simplify the production process, positively affecting the effectiveness of the entire enterprise [5]. One of the methods is Kanban, which belongs to pull systems, that is based on production to order not to the warehouse.

Figure 1 shows the overall material flow diagram for Kanban method.

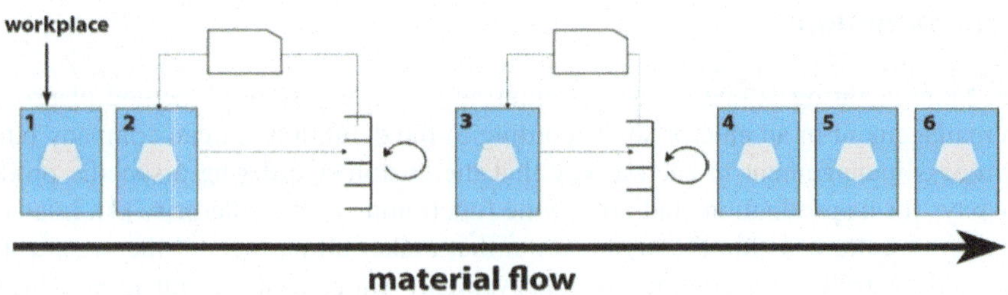

Fig. 1. General material flow diagram – Kanban. Source: own study.

Kanban consists in synchronizing the need to replace or supplement the materials necessary for production. It allows to integrate the material flow with the workplace for all workstations, reducing the amount of work-in-progress by using visual signals. In other words, it is about placing a work order or request for the delivery of a specific resource (raw, material, component, product, information) in a specific place, time, and quantity. Information can be sent in many ways. The simplest case is a signal that the product needs to be supplemented by the process, for example arrival of an empty container at a previous processing workstation. Another case is a card placed next to the goods during the production process, used to track current resources. Kanban can be used for a production order – it will allow to determine whether the production is ahead or late with the implementation of the schedule.

3 Drum-Buffer-Rope Method

Theory of Constraints focuses on the elimination or appropriate management of the constraints that occur in the flow. Goldratt proposes to remove or manage bottle-necks by using Drum-Buffer-Rope (DBR) methodology [6]. It presents its use to synchronize the use of resources and material flow in production operations [7].

Figure 2 shows the overall material flow diagram in DBR method.

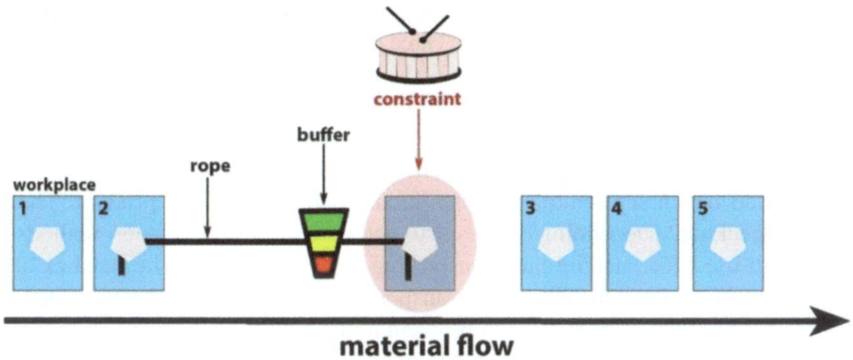

Fig. 2. General material flow diagram – DBR. Source: own study.

Each acronym in the BDR name has a special meaning:

(D) – The drum accentuates the pace at which the constraint works. Goldratt describes the bandwidth constraint as line drum beat. The drum is a constraint, a bottle-neck, a limitation. The rhythm of the drum indicates the maximum delivery rate because is the slowest.

(B) – The buffer determines the inventory at system or process checkpoints to protect against unforeseen changes. Prevents constraint inactivity caused by a lack of material to be processed. Ensures that the short interruption and fluctuations is no constraints do not contribute to the constraint. Traditional DBR has two buffers – one for the constraint and one for the shipment. The one against constraints is to protect the constraint, the shipping buffer protects the delivery date [8]. The dimensioning of the inventory buffers depends on customers' demand for a given product as soon as they are used up and how quickly they can be replenished [9]. The necessity to use buffers depends on how smooth the flow is (the more changes there are in the flow, the larger buffers should be used).

(R) – The rope provides communication between checkpoints to ensure system synchronization. This is the signal generated by the constraint, indicating that a portion of the inventory has been used up, releasing an inventory of the same size as the process. The role of the rope is to maintain capacity without creating excessive inventories [8]. The length of the rope, therefore, the amount of inventory in the system is determined by the protection against the constraint provided by the buffer. As the stocks of production in progress under the constraint are negligible, the rope works to maintain minimum and constant stock levels in the system [10].

4 Research Method

The research tools that were used: observation, interview, and comparative case study. A case study is a research methodology based on a detailed study of a single person, group, or event, especially to present, investigate the reasons underlying its principles. Presented as case-based science [11]. A case study was used in the analyzed enterprise. The purpose of the analysis was to identify bottlenecks in the process and to compare the application of the Kanban and DBR method. The packaging production process was analyzed. The processes occurring in the stream are planning, printing production, finishing processes, quality control, shipment to the customer. Three production workstations were analyzed successively in the process flow. The flow is the use of successive tasks associated with the production of a product without holding, scrapping, or moving backwards [12].

The first workplace is "punching", consisting in cutting sheets along the designed shape with a toll consisting of appropriately profiled blades. Next "stamping" workstation, carried out in semi-automatic presses, is used to finish the product. The last position is "packing". Depending on the customer's requirements, various methods and standards are used for packing and transporting a given product. The batches of the product are placed on pallets, which are transported to the storage bay, where they wait for the further production process. The problem noticed during the observation was the stock in progress, accumulating at the "stamping" workstation. The position had the highest average utilization, moreover 99% of the production was based on this resource. Therefore, the second workplace was indicated as a constraint of the process.

5 Case Study - Kanban

The production process is shown in Fig. 3 using the Kanban method in the analyzed enterprise.

In the Kanban method, the concept of Tact Time (T/T) appears, which determines the time in which the material moves through the entire production system. To match the customer's order, the production sequence must be in Tact Time. If the production time is shorter than the Tact Time, then the resources of the enterprise are not fully used, resulting in wasted waiting and an increase in the cost of production. The number of seconds per unit of product is the most common measure of the Tact Time [13]. In the analyzed case of the production process, the Tact Time is 0,4 s (seconds per piece). The product batch from the first workstation is made in a Tack Time equal to 0,25 s. The next workplace performs production with a Tact Time of 0,9 s. The second workstation performs the work the slowest, therefore it is defined as a bottleneck in the flow. So, the batch from the first workstation is transferred to the storage bay for the inventory

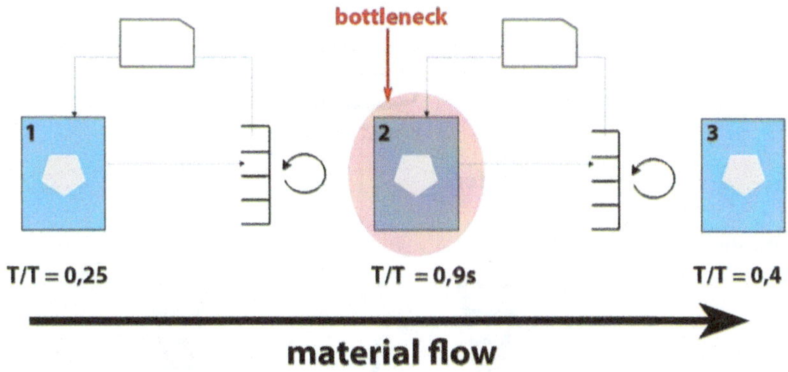

Fig. 3. Diagram of material flow in the analyzed production process - Kanban. Source: own study.

warehouse, which is called a supermarket in the Kanban method. The supermarket signals the quantity of each part needed from the upper supermarket - sending him Kanban in motion. Receiving a Kanban in motion causes the reloader to withdraw the required number of parts from the supermarket and then transfer these parts to the customer's resources. Creating a supermarket before the bottleneck requires creating a plan for the supermarket and filling inventories according to this plan (in this case the maximum available storage area). Marking a production order in green means, that the prepared materials and tools needed to perform the given order are complete. Marking an order in red indicates, that now there are not enough resources to complete the order. Thanks to this solution, operators notice the queue of prepared orders. If one order is incomplete, the operators in second workstation will execute the next order in the queue, minimizing the possibility of operator error. The completed production order is transferred to the next workstation in the flow, for which the tact time is 0,4 s. It can be noticed that the work on the third position is performed faster than on the previous position, therefore it is waiting for the next batch of the product. That's why, to minimize the wastage arising, position three works in two shifts. The product batches from the second workplace are placed in the supermarket in front of the third workplace during the downtime.

6 Application DBR Method Instead of Kanban Method

The applications of the DBR method instead of the functioning Kanban method in the analyzed production process are presented in Fig. 4.

According to the idea of the Theory of Constraints, all resources should be focused on the identified constraint [7]. In the analyzed case, the constraint is the "stamping" workplace. All workstations work at the constraint rate (D), the tact time is 0,9 s. To prevent unlimited inventory build-up ahead of the slowest workplace, so material should be "released" at a constraint rate. The rope (R) acts as a final for inventory release in the process, ensuring process throughput. Each time the constraint ends a unit of work-in-progress (WIP), the rope beeps (for example in the form of a message in the integrated management program) to release the raw material unit to the buffer at the level of workstation one. Before constraint, buffer (B) was placed. If the first workstation is

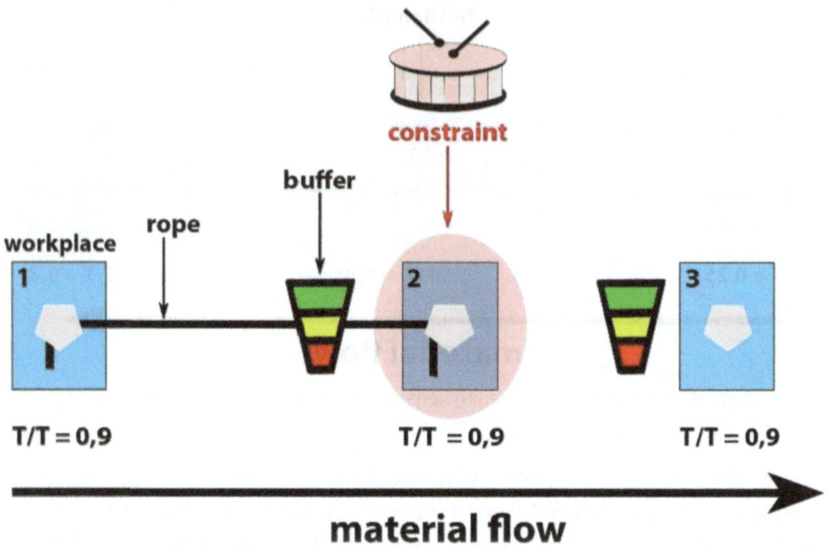

Fig. 4. Diagram of material flow in the analyzed production process after applying DBR. Source: own study.

in a failure state, the (WIP) units are placed in the buffer [14]. An important step is to determine the size of the buffer that should be adjusted to the constraint. The buffer may be presented as the time that work-in-progress should be delivered before being used first. Referring to the case study, the buffer may be 1 h, therefore approximately 4,000 batch pieces should be delivered to the constraint site every hour. Such action ensures the stability of the stock operation. The greater the variety of the process, the greater the buffers must be. The stock buffer can be divided into 3 areas, it is marked by visual criteria: top green level, yellow and red. The first green means that the stock is sufficient, and no action is required. If the buffer status is yellow, it means that the order should be shipped. In case the buffer level reaches the red level, it means that replenishment plans must be executed immediately. In the analyzed case, the safe stock may cover the range of 4000–2001 items, the alarming level 2000-501 items, and the critical level 500-0 items of packaging. By monitoring the three levels shown, for each order over a period, it will be possible to determine whether the buffers are approximately real in relation to the consumption rhythm [9]. Planning of the production schedule is based on the productivity data of key workstations and working time in buffers. There is no need to collect detailed information from all workstations, because in the theory of limitations in the production process, the operating status in buffers before bottlenecks is controlled. Information on what to produce, in what order and quantity is determined by the schedule on the drum (D).

7 Discussion

Both Lean and TOC focus on continuous improvement and control of material flow in the production hall. Their use allows you to significantly simplify operations, increase the implementation time, and the profitability of the enterprise. Correctly implemented DBR with TOC results in resource synchronization and effective use of materials in relation to the identified system limitations. Both methods are aimed at more efficient use of resources and improvement of operational efficiency [15]. Each of the methods does it in a different way [16].

Table 1 shows a comparison between Kanban and Drum-Buffer-Rope.

Table 1. Comparison between Kanban and DBR method.

Area	Comparison methods	
	Kanban	Drum-Buffer-Rope
Idea	Increasing profit by enhancing the added value of the product from the customer's point of view	Increase your profit by increasing your throughput
Organization of the flow	Continuous	Compliance with constraint
Stimulation	The customer sets the pace	Constraint sets the pace
Production enviroment	Repeatable production	Serial and unit production
Flow type	Pull	Pull
The sequence of operations	FIFO	FIFO
Methods implementation steps	5 Lean principles	POOGI
Inventory	Elimination of all possible	Handling enough quantity to maximize the flow of constraint
Component	Supermarket, Kanban	Buffer, rope
Result	Lower production costs and prevention of waste	Increased production capacity

Source: own study.

Kanban method and DBR strongly focus on the customer, as a result companies achieve favorable, efficient process management. When comparing Kanban with DBR, the following similarities can be noticed:

1. Both cases are pull systems (as opposed to the "push" method where long series are created, based on previous sales forecasts). In a trailed system, successive processes define the production needs in the previous processes. They only "take out" the goods they need, in the time and in the quantity, they need [17]. They synchronize the productive "activity" of the process with each other on the higher and lower streams [13].
2. The sequence of operations in both methods is based on FIFO ("First-In, First-Out"), which is a method in which goods from the warehouse are issued in the order in which they were entered into this warehouse. FIFO helps to manage the

assortment in the warehouse in an orderly and efficient manner. Avoidance of expiry of the materials and wastage of making complicated calculations of the valuation of individual assortments over time [18].
3. The best way to shorten batch transit times is to eliminate stockpiling - nothing stops work, production starts to "sound" as it should. The Kanban method prevents overproduction and excessive movement of materials between processes. It provides actual production orders to processes according to the principle of replenishment, by synchronizing the time of movement of materials and the amount of delivered materials. It supervises the production by means of visualization, which shows whether the production is delayed or ahead of the schedule. Correctly used Kanban will prevent you from overloading multitasking. Similarly, DBR minimizes the excessive flow of materials because it adjusts their quantities to the pace of work of the constraint, considers the problem of waste of resources [19].
4. DBR in Theory of Constraints is like the Kanban from the supermarket before the bottleneck. Every portion is taken from the buffer/supermarket, the information is sent via the rope/Kanban to the beginning of the rope/Kanban loop to replenish the materials [20].

The following information can be distinguished among the differences:

1. The stages of implementing changes in Kanban are based on the concept of Lean management. In DBR, the stages of implementing changes focus on 5 steps of POOGI (Process of On-Going Improvement): identifying the constraint, exploiting the constraint, subordinating everything to the constraint, increasing the power of the constraint, returning to the beginning.
2. Unlike DBR, Kanban offers guidance on what to produce next - waiting in the Kanban queue. This possibility minimizes the risk of error.
3. In Kanban, if one workstation stops working for longer than the buffer allows, the whole process is forced to wait until the suspended workstation is restored. In the case of DBR, the risk of production stoppage exists when the constraint is unable to process resources due to failures, shortages. In DBR, it is not possible to make up for losses in the event of losing any production on the drum [7]. Therefore, all possible measures must be taken in DBR to keep the drum running continuously and efficiently. Increasing the productivity of the drum will lead to an overall increase in process efficiency [21].
4. The Kanban method to be effective, requires a stable production environment and uniformity of production [22]. It is most effective in a production environment with regular and constant demand, where products have a simple and flat bill of materials, short lead times and low order volumes [23]. DBR can be used in a production environment based on both serial production and unit production [24].
5. The use of DBR makes sense to standardize the process as it will indicate where to improve to have the greatest impact on performance and throughput. It prevents system overload as it minimizes work-in-progress. Keeping your WIP to a minimum means optimizing your flow and lead times. It leads to systematic and sustained process improvement that Kanban is not capable of [25].

6. The criterion for applying Kanban is the occurrence of a signal, usually visual, limiting WIP. In Kanban, the point of commitment is important, symmetrical between the business and the enterprise providing specific services. The criterion for applying DBR is the presence of a constraint in the process.

DBR technique - was developed to meet the basic assumptions of the Theory of Constraints, aimed at streamlining production processes to increase throughput and reduce inventory and operating costs [26]. Kanban visualizes the workflow, defines work-in-progress limits and the measurement of lead times. It enables a proactive approach that solves the problems of downtime and bottlenecks when a signal occurs, not after a noncompliance. Kanban is a universal solution [27].

The main factor that causes the Theory of Constraints not to be widely applied is the need to change the financial paradigm. It requires enterprises to „give up" what they know, that is the classical thinking process. Enterprises want to make the most profit in the shortest possible time, not necessarily slowing down the machines. It is a classic approach to production management, which is focused on working with 100% use of production capacity. The Kanban method is widely and willingly used by enterprises. Of course, the Kanban mantra is to start the next steps when the work in progress is completed. However, improper use, not controlling Kanban, involves a hidden, excessive amount of work-in-progress. Determining the size of the buffer in an environment with high uncertainty in the supply chain is very important, however, in many companies in practice it is determined by trial and error. Using the right approach to determining the size of the buffer, the level of WIP units in the supply chain would remain adequate, with minimum inventory cost and maximum response to customer demand [28].

To increase the throughput of the production process, enterprises invest in the development of a position which is a constraint or use cooperation consisting in delegating part of the production to the bottleneck. Using outsourcing services is beneficial because it allows you to achieve a certain level of sales while maintaining low production costs, however, it also involves the risk of patent leakage or the company's dependence on an outsourcing company [29].

Nowadays, most companies decide to introduce agile management methods, for example Agile rather than TOC, for some projects the cost of introducing the critical chain is too high. Enterprises consider the implementation of TOC as economically unprofitable, as they require a change in the company's organizational culture, training of employees and fighting their old habits, or purchasing software [30]. Adjusting people's work to the constraint is often the most difficult stage. During the adaptation phase, resistance from workers who are used to their work based on performance indicators is encountered [31]. This approach may result in excess inventory and a shift in priorities in various positions that will not coincide with the company's goals and the use of DBR. The development of robotization and production automation may lead to changes in the culture of enterprises employing good and open managers who will be willing to implement DBR. Synchronization of the speed of the feeders and robots to be reduced may be easier to implement and maintain (for example in economic terms).

8 Conclusions

A one-piece flow is ideal to detect any waste in your production processes. Production without stocks, except in exceptional circumstances, is practically impossible, and the appropriate methods of the suction system allow you to determine their correct level [13]. The continual improvement process in the Theory of Constraints is a consequence of concentrating all efforts on the purpose of the system. It must not be possible to lead to a situation where inertia (concentration on a resource which, from the point of view of the entire system, is no longer a constraint) becomes an obstacle, a constraint of the system [7]. The elimination of one constraint of the system does not exclude the appearance of another. Each process is variable; therefore, it should be constantly analyzed and the elements (drum position, rope size, buffer size) should be constantly adjusted [20].

The mere establishment of a permanent bottleneck will lead to problems. The greatest similarity between the DBR technique and Kanban is the minimization of overproduction and excessive movement of materials between processes. Kanban can fulfill DBR roles. An example is the occurrence of work-in-progress limits, in which the production adjusts to the work pace of the process, new production orders are not "released" until the open tasks are completed. In comparison, this is the same DBR application case where the environment adjusts to the constraint. Both methods provide the actual production orders to the processes according to the principle of replenishment, by synchronizing the time of movement of materials and the amount of delivered materials. They supervise the production by means of visualization, for example in buffers, which shows whether the production is delayed or ahead of the schedule. Correctly used Kanban will prevent you from overloading multitasking, adjusts their quantity to the pace of work of the constraint, considers the problem of waste of resources [19]. It is possible to combine these methods, for example in production environments that execute several orders simultaneously [32]. Supporting them by computer will positively affect the streamlining of the flow, eliminate multitasking reducing the efficiency of the entire enterprise [33].

Subsequent studies of the Kanban and DBR comparisons may include a broader analysis beyond simple serial flow lines with more than 3 stations. It is also possible to compare a computer simulation of the application of the DBR method in a selected company process, in which Kanban is currently used.

If a satisfactory simulation is obtained, the tests can be applied to the actual implementation of DBR. Simulation studies may be carried out on more than one enterprise to confirm the research problem. Interviews with production managers, Lean specialists on the application of the Kanban method and DBR by companies from various industries will allow to obtain key practical knowledge that will allow for a more accurate reflection of the actual process in a simulation study.

Acknowledgments. This research was funded by a grant from the Ministry of Education and Science received by the Bialystok University of Technology, grant number WZ/WIZ-INZ/3/2019.

References

1. Szczepański, M.: Lean management - automation as a tool to improve the competitiveness of enterprises on the production services market, Sophia, pp. 98–110 (2016)
2. Monden, Y.: Toyota Production System: An Integrated Approach to Just-in-Time, 2nd edn. Industrial Engineering and Management Press (1993)
3. Lizarralde-Aiastui, A., Apaolaza-Perez de Eulate, U., Mediavilla-Guisasola, M.: A strategic approach for bottleneck identification in make-to-order environments: a Drum-Buffer-Rope action research based case study. J. Ind. Eng. Manag. **13**(1), 8–37 (2020)
4. Ching, C.: The Bottleneck Rules: How to Get More Done (When Working Harder isn't Working). Independently Published (2018)
5. Woeppel, M.J.: Manufacturers Guide to Implementing Theory of Constraints. The CRC Press Series on Constraints Management (2000)
6. Goldratt, E.M., Fox, R.E.: The Race. North River Press, New York (1986)
7. Goldratt, E.M.: The Goal: A Process of Ongoing Improvement. North River Press, Great Barrington (1984)
8. Lang, L.: Drum Buffer Rope (DBR) Summary. https://www.velocityschedulingsystem.com/blog/drum-buffer-rope/. Accessed 05 July 2021
9. Prostean, G., Badea, A.: Supply chain collaboration model based on Drum-Buffer-Rope philosophy. In: Balas, V.E., Jain, L.C., Kovačević, B. (eds.) Soft Computing Applications. AISC, vol. 357, pp. 1025–1035. Springer, Cham (2016). https://doi.org/10.1007/978-3-319-18416-6_82
10. Mohammadi, A., Eneyo, E.S.: Application of drum-buffer-rope methodology in scheduling of healthcare system. Architectual Engineering – Faculty Publications (2012)
11. Yin, R.K.: Case study research. In: Joanna, G. (ed.) Kraków Publishing House of the Jagiellonian University (2015)
12. The Productivity Press Development Team: Pull Production for the Shopfloor, Wrocław (2011)
13. Smalley, A.: Creating Level Pull. Lean Enterprise Institute Polska (2011)
14. Betterton, C.E., Cox, J.F., III.: Espoused drum-buffer-rope flow control in serial lines: a comparative study of simulation models. Int. J. Prod. Econ. **117**(1), 66–79 (2009)
15. Forte, T.: Theory of Constraints 101 (2016). https://fortelabs.co/blog/series/toc-101/. Accessed 05 July 2021
16. Łopatowska, J.: The use of the theory of constraints to the implementation of the principles of lean. Mater. Manag. Logist. **11**(CD), 49–51 (2014)
17. The Productivity Press Development Team: Kanban for the Shopfloor, Wrocław (2009)
18. Walentynowicz, P., Bierzanowski, R.: Using the LEAN concept in logistics on the example of the C-L distribution company. Studia Informatica Pomerania **1**(43), 91–103 (2017)
19. Hetmańczyk, H.H.: Theory of Constraints in the context widely understood educational activities. Colloquium WNHiS AMW Quarterly I (2013)
20. Roser, Ch.: All About Pull Production: Designing, Implementing, and Maintaining Kanban, CONWIP, and Other Pull Systems in Lean Production. AllAboutLean (2021)
21. Koliński, A.: Using the analysis of bottlenecks in production management. Mater. Manag. Logist. **9**, 16–21 (2010)
22. Ohno, T.: Toyota Production System: Beyond Large-Scale Production. English translation by Productivity Press, New York (1988)
23. Clotet Joaquim, F.: Lean production planning and control in semi-process industries. Norwegian University of Science and Technology (2015)
24. Schmidt, P.: "What is the Theory of Constraints?", TOC + Constraints Management Institute (2020)

25. Tendon, S.: Enhanced risk management in Kanban via the theory of constraints. In: Lean Kanban Netherlands Conference (2012)
26. Cox III, J.F., Schleider Jr, J.G.: Theory of Constraints Handbook. McGraw-Hill Education (2010)
27. Szwedka, K., Lubiński, P.: Kanban system implementation concept - case study, innovations in management and production engineering. Publishing House of the Polish Society of Production Management, Opole (2015)
28. Parsaei, Z., Nahavandi, N., Elmekkawy, T.: Buffer size determination for drum-buffer-rope controlled supply chain networks. Int. J. Agile Syst. Manag. **5**(2), 151–163 (2012)
29. Koliński, A.: Elimination of bottlenecks as a tool for optimization production. Mater. Manag. Logist. **10**, 20–26 (2010)
30. Kabut, M.: The theory of constraints - concept and practical examples. WZ UW (2011)
31. Rzepa, M.: The theory of constraints or TOC. A modern warehouse. Mag. Storage Warehous. Syst. **10**(6), 71–74 (2008)
32. Moroz, E.: Application of theory of constraints in the implementation of computer management systems for the production process. Mechanic **7**, 772–773 (2016)
33. Dinur, Y.: DBR Kanban - Overcoming a boundary of DBR application (2016). https://www.linkedin.com/pulse/dbr-kanban-overcoming-boundary-application-yaniv-dinur. Accessed 28 June 2021

A Prediction Approach for Aluminum Extrusion Processing Using Neuro-Fuzzy Based Decision Making

Suthep Butdee(✉)

King Mongkut's University of Technology North Bangkok, 1/7 Soi Noparat 1/3 Krisadanakorn 20 Taweewatana, Bangkok 10170, Thailand
suthep.butdee@gmail.com

Abstract. This paper proposes a prediction approach for aluminum extrusion processing using Neuro-fuzzy based decision making under multi-criteria variables. The ANFIS system model is developed with 4 inputs and one output membership functions that are designed by 5 levels. The fuzzy inference rules of the extrusion domain knowledge are captured from experts and documentations. Four factors that are influenced the quality of the extrusion profile are concerned; extrudability, extrusion ratio, hydraulic pressure and ram speed. The ANFIS system consists of 625 rules. The model is simulated by using MATLAB toolbox with a case study. Validation is then tested by 3D simulation using Hyperxtrude software. The result is headed the same direction with the ANFIS prediction. The contribution of the paper is the development of ANFIS extrusion processing model for prediction quality of a profile at the design stage. The major benefit is to assist design engineer to make efficient decision and reduce risky, reduce errors and defects of the extrusion processing.

Keywords: A prediction approach · Aluminum extrusion processing · Neuro-fuzzy based decision making

1 Introduction

Aluminum extrusion is a hot deformation process which is widely used to produce long straight aluminum profiles such as bars, solid, hollow, tubes and so on. A hot billet is pressed under high pressure and temperature in a specific machine. A billet is pressed throughout a die to form a profile section [1]. A profile shape is deformed according to the die orifice shape of the die design. Therefor, the die needs to be strong sufficiently for resisting high pressure from hydraulic ram of a machine. A hot billet is crushed through the die which is preheated at approximately 450 °C [2]. There are two main types of the die; solid and hollow die. The solid die consists of four components. They are the die plate, backer plate, feeder plate and bolster plate. The hollow die consists of four components. They are mandrel, die cap, bolster and sub-bolster. Die plate is activated as the major components composed of the hole of profile shape including orifice hole using for reducing friction. The die plate must be designed to obtain a strengthen against the

A. Hamrol et al. (Eds.): MANUFACTURING 2022, LNME, pp. 237–249, 2022.
https://doi.org/10.1007/978-3-031-00218-2_20

high pressure of the hot billet extrusion. Backer plate is used to support the tongue of the die plate to prevent collapse and distortion. In addition, the backer plate composes of backer orifice which is related to the die orifice. Feeder plate is used to create balance flow of the hot billet through the die orifice. Bolster plate is used to support extrusion load which is transmitted from the die and backer. Die holder holds the die plate whereas the die carrier holds the die set in the extrusion machine. Bridge is used to divide the hole of billet flow and support the mandrel. Even through the die is perfectly designed and the billet is well prepared, the quality of the profile is still relied on the capability of the machine and the efficiency of controlling on various parameters such as pressure and temperature.

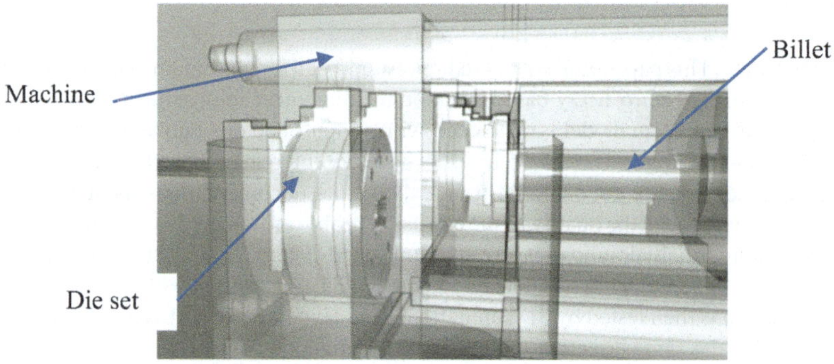

Fig. 1. Aluminum billet, die set and aluminum profile [3]

Figure 1 shows the three main parts of aluminum extrusion processing which affect to the profile quality. They are a die-set, a machine and a billet. In addition, some errors from the processing control together with defect prevention need to be studied beforehand. The trouble shooting and defect prevention are used and created as fuzzy inference rules inside the ANFIS model. The defects are die-stop, chatter, broken surface, hot tearing, die lines, blisters, laminations, weld line, structure streaks, orange peel, waves, twist, kinks, wire edge, shape edge, stains, rubs, dents and gouges. There a lot of evidences for adopting Neuro-fuzzy system to different domains in manufacturing production. The details are explained in the next session. This paper therefore proposes a prediction approach for aluminum extrusion processing with multi criteria decision making. The model is implemented on the MATLAB platform and validated with 3D simulation on Hyperxtrude software. The knowledge is based on the information collected from the aluminum extrusion factory in Thailand.

2 Literature Reviews

Aluminum extrusion processing is a complex production process and uncertain circumstance due to a hot aluminum billet is pressed and crashed by a high-pressure force and temperature inside a strong cylinder of a press machine throughout a small area of die in different and various shapes. In general, there are three major factors that impact the extrusion process; machine, material and die. Therefore, the machine control parameters need to be controlled effectively. Material preparation should be controlled correctly and sufficiently particularly when the material's property changes, the preparation must be also changed according to their characteristics. Die is one of the most critical and complicated as well as sensitive to affect the quality of the profile (finished goods). A solid die is quite simple whereas hollow die is more complicated. Die design is then widely concerned by both in academic study and in practice. Hollow die is quite complicated. The hot billet is split inside the hollow die and flowed around a mandrel to form a hollow profile. The quality needed from the process is to join a seam weld perfectly and same strength as the bulk structure. The simple example is an aluminum round tube.

Extrusion processing is sensitive because of high temperature as well as high pressure control. The process consists of log casting, homogenization, preheating, extrusion, stretching and aging. The log casting involves with a molten metal of billet which a combination of prime alloy and recycled metal. It is treated to become a nonuniform microstructure which leads increase strength, grain structure and ductility. Homogenization is a process to transform specific phases with suitable time and temperature. The billet is preheated by gas and induction. Machine factors are also needed to control effectively as extrusion parameters such as ram speed, cylinder temperature control with different stages, pressure control and so on. The machine pressure control is ranged from 500 to 11,000 ton for billet diameters from 3″ to 24″. The next process is quenching and stretching as well as anodizing. One of the output efficiencies is measurement by the profile weight. In general, yield of production is approximately 63%–80% [1].

This section presents the previous research articles involving with aluminum extrusion processing, quality prediction, research tools, control parameters such as temperature, pressure, ram speed, and defects including their solutions in order to used for creating fuzzy inference rules in the ANFIS model. Previous researches have been done to achieve optimization processes and machine setting parameters such as Kathirgamanatha and Neitzert [4] proposes optimization method of extrusion dies using FEM. Due to the aluminum extrusion process is complex and uncertain with several effected factors, therefore prediction is needed to reduce risk and failure. Optimization study is on die geometry and friction. Zhao et al. [5] proposes process modeling and die optimization design of aluminum alloy extrusion profile used in high speed train based on multi-objective method. Kapadia and Desai [6] studies on optimization of die extrusion process using FEM method. The objective is to obtain defect minimization as well as improve quality and productivity. The simulation is on Hyperxtrude under the condition of AA6063 material. The billet size is on 150 mm. and preheated by 450–500 °C. The ram speed is varied between 40–60 mm/s. Pressure is set approximately between 470 to 760 Mpa. Park et al. [7] presents evaluation of the effect of ram speed for extrusion of AA6063 based on ALE-based finite element analysis of L-shapes sample. In terms of productivity, selected papers are shown such as Ferras et al. [8] presents scrap production of extruded aluminum alloys by direct extrusion. This study is in Portuguese company. Aluminum extrusion is a process that requires appropriate processing parameters to be used in order to produce diverse profiles and high-quality products. Butdee [9] proposed adaptive aluminum extrusion die design using case based reasoning and artificial neural networks. Instead of design a new die from scratch, this method surveys and searches previous used dies which are similar to the new die design requirements and then adapts and modifies one of them. This method is known as case-based reasoning. Due to the retrieval die cases in the databases or case library which is time consuming and inefficient, ANN approach is adopted and developed to deal with retrieval case problems. Feature-based die (aluminum profile shape) model is the main indexed together with critical design parameters such as bearing length, thickness, tongue ratio, dimension, cross section area, extrusion ratio, circumstance circle diameter (CCD). The outputs are billet length, quality of profile and yield. Then, there are some researches that has been done on optimization process and machine control parameters.

Zhang et el. [10] presents fatigue failure prediction model and verification of hot extrusion die. The fatigue properties of materials are closely related to service conditions and heat treatment process. Negozio et al. [11] explains validation process using FEM for front end and back end defects evolution in AA6063 and AA6082 aluminum alloys profiles. The simulation is performed by HyperXtrude software. Yadav et al. [12] proposes finite element analysis of extrusion process using aluminum alloy whereas Fang et al. [13] presents FEM simulation of aluminum extrusion through two-hole multi-step pocket dies which is used for a solid aluminum profile. Kniazkin and Vlasov [14] proposes quality prediction of longitudinal seam welds in aluminum profile extrusion based on simulation. Vallberg et al. [15] presents the mechanism of formation of back-end defects in the extrusion process using FEM which is applied to study metal flow in aluminum extrusion focusing on how material flow defects from when the last part of the billet is extruded through an axisymmetric die. Then, the software DEFORM is used to model the extrusion process. As mentioned from the previous researches, data can be collected as case study combining with standard data from the aluminum extrusion technology and reformulated to develop a new model used for prediction. Intelligent systems have developed using various methods such as fuzzy logic [16], Artificial Neural Networks [17], Neuro Fuzzy system, Fuzzy AHP [18] and so on. In addition, some researches are employed numerical simulation, DEFORM, and Hypertextured simulation software. Modern research studies intelligent methods such as fuzzy logic, artificial neural network, neuro-fuzzy system, and fuzzy AHP for deciding multi-criteria decision on complex and uncertain environment.

3 Research Methodology

This section explains the research method used in this research. ANFIS model is developed which is used the concept of neuro-fuzzy intelligent system. The ANFIS rules are created inside the MATLAB toolbox. The input membership functions are also created together with the output membership function. The neuro-fuzzy system is a hybrid and a combined model of neural network and fuzzy logic. It consists of 4 layers: fuzzification, fuzzy rule, normalization and defuzzification. The ANFIS model can lean both forward and backward, known as backpropagation, which is used to compute the error of the neurons in the hidden layers.

The neuro-fuzzy approach is adopted to deal with uncertainty and complex process of aluminum extrusion at the beginning of the process life cycle that is an effective modelling simulation. The objective is to predict quality of the profile when critical variables are given to the model. There are three major factors and four variables which impact to the quality profile; machine parameters, die geometry and billet material. Four variables of extrudability are considered for the billet preparation's capability. Extrusion ratio is selected to be taken into account for the quality of the die design. If the extrusion ratio is properly chosen, the high-quality profile is carried out. Two more parameters are concerned represented the ability of machine controlling. They are hydraulic ram pressure and ram speed. In addition, the assumption of the simulation is considered based on the material AA6063, solid die, billet size is 16 inches, machine capability is 600 ton. Neuro-fuzzy model is created by using the MATLAB tool box platform whereas the extrusion modelling is simulated by Hyperxtrude software,

The next step of the data preparation for creating the neuro-fuzzy approach which is always called ANFIS model is to collect the physical parameters of the extrusion processing in order to assign the fuzzy membership functions. Billet is a critical factor for the extrusion processing which is controlled by the temperature compensated strain rate. It should be more than 20. The tension limits for attainable strain is 0.3 that can increase up to 2.0 when the lubricant is applied. Neuro Fuzzy model is developed based on the extrusion processing performance evaluation related to die design factors. Figure 2 shows the ANFIS model development for aluminum profile quality from extrusion processing. The are 5 criteria are concerned such as billet conditions which need to control mainly on temperature. Die geometry consists of many values but extrusion ratio is the major impact to quality of the processing output whereas hydraulic pressure is a major power force that pushes the billet with high pressure. Extruding speed relates to the quality of the profile. The speed relates to the material flow, stress and strain of the profile. ANFIS system consists of fuzzy rules and inference reasoning to conclude the results of questions and solutions. The goal of the model is to predict quality of aluminum extruded profile.

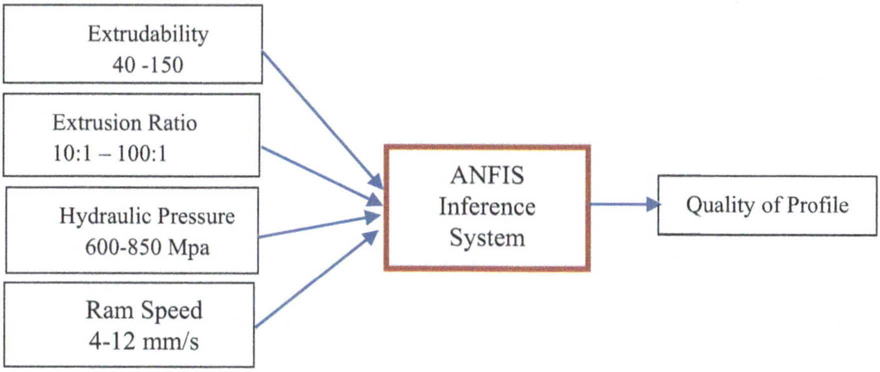

Fig. 2. ANFIS model development for the aluminum profile quality prediction

One of the critical factors of the extrusion processing is material billet. The billet is heated at the suitable temperature. However, there three sources of heat effected to the billet processing in the extrusion process which is called heat balance. Heat balance for extrusion process is significant to influence the quality of the profile extrusion. Heat balance combines billet heat, heat of deformation and heat losses in container. The heat losses are based on the effect of cooling and container zone offsets. The heat is multiplied by mass, c_p and Δt. Therefore, the total heat is the combination of billet heat and deformation heat which is subtracted by the heat losses and become the extrusion heat.

Table 1. Extrudability of aluminum alloys [2]

Alloy	Degree	Alloy	Degree
1060	150	6066	40
1150	150	6101	100
2011	150	6253	70
3003	100	6351	80
5083	80	6463	60
6061	60	6663	100
6063	100	7078	7

Table 1 shows extrudability of aluminum alloys. It represents the overall capability of the billet quality that can produce the final profile according to the design specification. It includes billet temperature, billet types and property. However, this paper does not involve with the detail but concerns only the extrudability. From Table 1, it can be divided the fuzzy membership functions into 5 levels based the extrudability degree. Very High (VH), High (H), Medium (M), Low (L) and Very Low (VL). The range of the extrudability is defined as follow.

- VH is between >100 to 150.
- H is between 80 to 100.
- M is between 60–79.
- L is between 40–59.
- VL is <40.

The second criteria influenced the profile quality is extrusion ratio which is the proportion of initial cross-sectional area (Dc) and final cross-sectional area (De). Extrusion ratio (ER) and tongue ratio (TR) are influenced by die which is a critical criterion to the extrusion processing. It is the comparison of billet area and profile area. Tongue area is the proportional area between width of the gap and its width. Both extrusion ratio and tongue ratio cause the difficulty of extrusion processing. The higher ratio occurs, the greater pressure is needed to force the metal through the smaller opening die. In addition, some parameters effect to the profile quality such as the factor and the tolerance. The factor is determined by the perimeter of shape and weight per meter. Tolerance is a limitation of variation in which a profile can be produced related to extrudability. However, if the Er of the profile section is low, then the parts of the shape will be less mechanical work performed. The acceptable Er range is between 10:1 to 35:1 for hard alloys whereas the Er of the soft alloys range between 10:1 to 100:1 or up to 400:1 for aluminum extrusion. As the result, the higher of the extrusion ratio, the harder is occurred. The extrusion ratio is De/Dc which is shown in the Fig. 3.

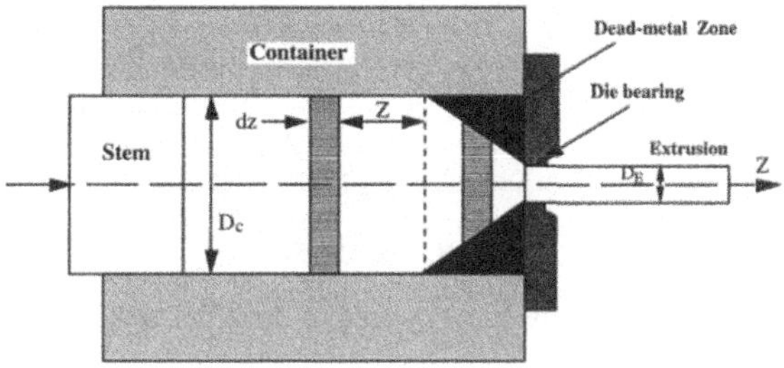

Fig. 3. Extrusion ratio and dead metal zone [2]

The third criteria are hydraulic ram force. The extrusion force can be calculated by the Eq. 1 and Eq. 2. The working force is depended on material type and material property including temperature, the diameter of billet.

$$\overline{\sigma}\int\frac{\bar{\mathrm{d}}L}{L}=\overline{\sigma}\ \ln\frac{L_2}{L_1}=\overline{\sigma}\ \ln\frac{A_2}{A_1}=\overline{\sigma}\ \ln\ R \tag{1}$$

$$F=AokIn\left(\frac{Ao}{Af}\right) \tag{2}$$

where:

k = extrusion constant
Ao, Af = billet and extruded product area

The fourth criteria are ram speed which is concerned in this model. The ram speed is related to billet temperature. Figure 4 shows the diagram of the ram speed zone for saving extrusion. If the ram speed increases, then the pressure load will increase. On the other hand, the temperature decreases when the ram speed increases.

Extrusion ram speed effects to the mechanical property of the profile which gives higher engineering stress and longer engineering strain. Normally, the ram speed is set between 4–12 mm/s. The extrusion pressure increases when the ram speed increases whereas low ram speed leads to cool the billet. As a result, the flow stress is increase. ANFIS system is used to make decision based on rule-based information which is created from the combination of the 4 inputs and 1 output. Rules are captured from hand book and from the factory experts.

Fuzzy rules are created based on the experts related to the input membership functions and the output membership functions. The objective function is to predict quality of extrusion processing profile. From the fuzzy rules, they are transformed into the system. Partition of the result generated from the ANFIS model is shown in the Table 2.

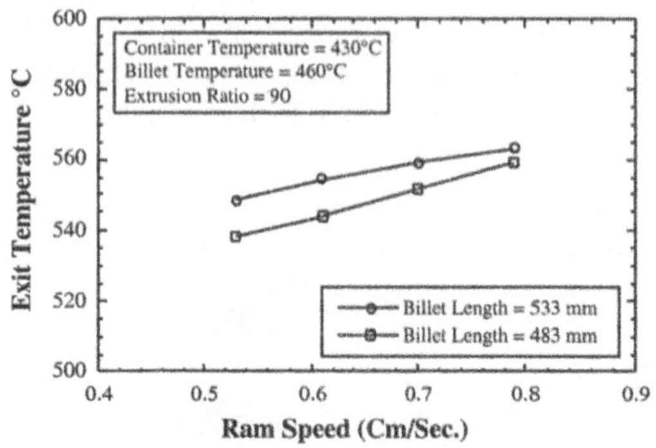

Fig. 4. Extrusion ram speed of the billet heated

Table 2. Partitions of the ANFIS model modeling

No.	Extrudability	Extrusion ratio	Hydraulic pressure (MPa)	Ram speed (mm/s)	Quality profile prediction by experts	Model Prediction by ANFIS
1	40	10	600	30	75	75.55
2	40	10	600	40	76	76.69
3	40	10	600	60	77	77.61
4	40	10	600	80	78	81.69
5	40	10	600	100	79	80.90
6	40	10	700	30	75	74.99
7	40	10	700	40	76	75.13
8	40	10	700	60	77	76.69
9	40	10	700	80	78	77.45
10	40	10	700	100	79	78.89
11	40	10	750	30	76	75.84

4 Simulation Modelling

This section presents the simulation modeling of the ANFIS model based on the methodology in the Sect. 3. There are 4 input membership functions; extrudability, die extrusion ratio, hydraulic ram pressure, and extruding speed as shown in the Fig. 5. The ANFIS model consists of inference engine using fuzzy rules linked to the output membership function which aims to predict quality of the profile. The ANFIS is created based on Sugeno approach.

The first factor of input variable is billet conditions which concerns with extrudability. This input is designed by 5 levels of membership function as shown in the Fig. 6. The membership function is referred to the data of extrudability degree. Aluminum extrudability is ranged between 40 to 120. The triangle fuzzy approach is selected.

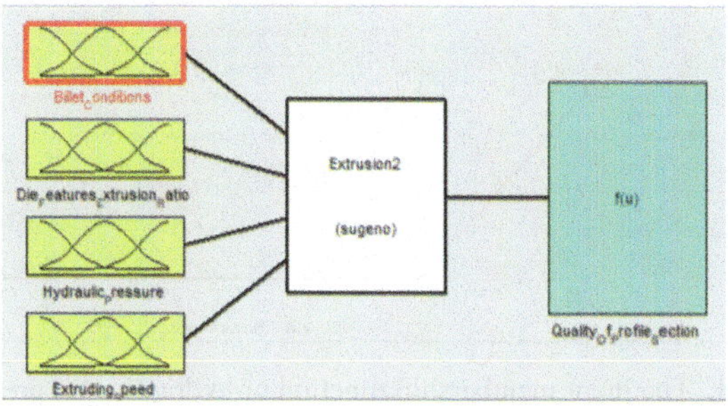

Fig. 5. ANFIS membership functions of the quality profile prediction

There are three points that are concerned; the base point on the left and the right of the triangle, and the top point. For example, medium extrudability ranges between 60 to 100 but the mid-point is 80.

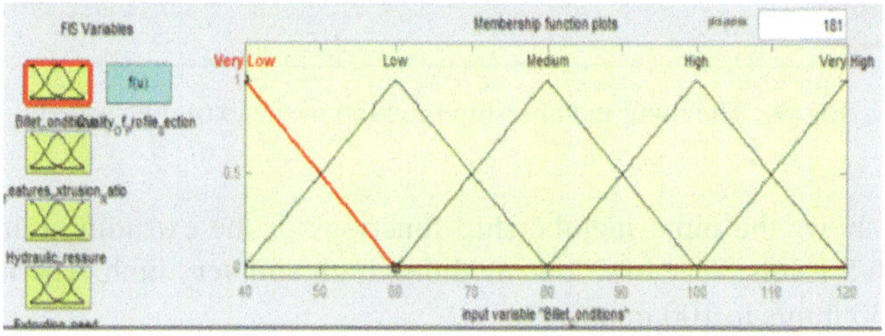

Fig. 6. The input membership function of the extrudability of the billet conditions

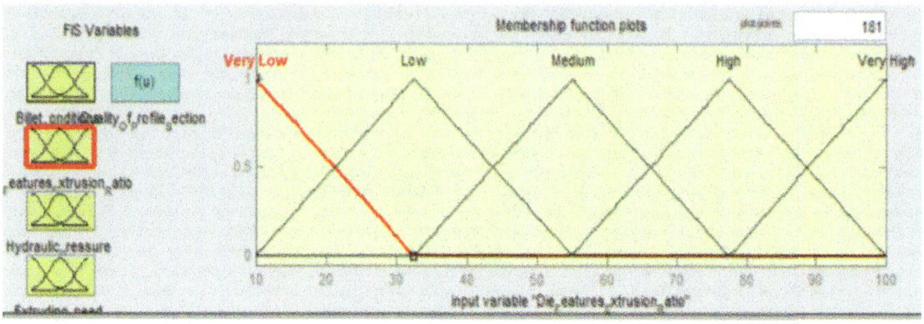

Fig. 7. The input membership function of the die extrusion ratio (ER)

Figure 7 shows the input membership function of the die extrusion ratio. The model is designed to create the membership function in 5 levels; very low, low, medium, high and very high. The ER range is between 10 to 100.

Figure 8 shows the input membership function of the hydraulic ram pressure. It is divided into 5 levels; very low, low, medium, high and very high. The pressure range is between 600 MPa to 850 MPa.

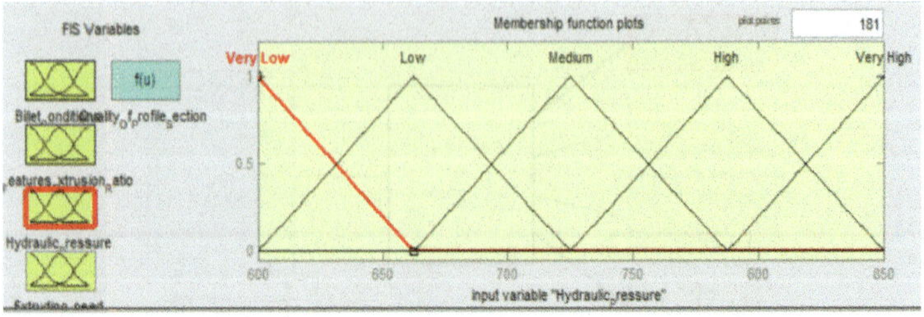

Fig. 8. The input membership function of hydraulic ram pressure

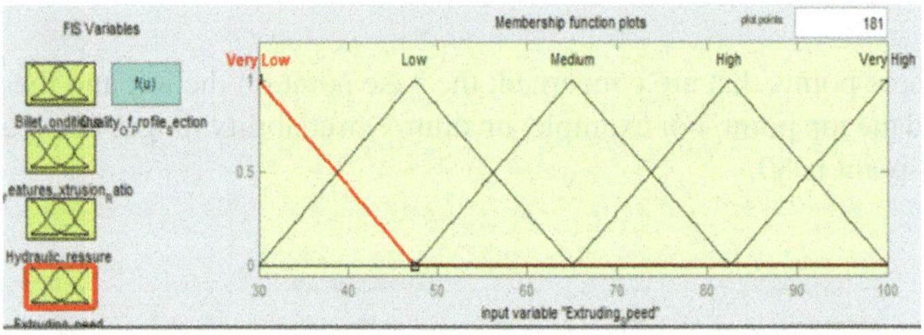

Fig. 9. The input membership function of the extruding speed

Figure 9 shows the input membership functions of the extruding ram speed. It is designed with 5 levels; very low, low, medium, high and very high. The input variable ranges from 30 mm/s to 100 mm/s.

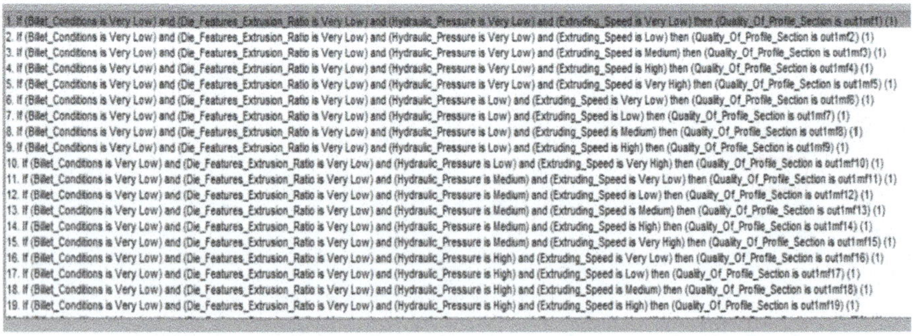

Fig. 10. The fuzzy rule created in the ANFIS inference system

Figure 10 shows the fuzzy rule-based decision making for aluminum extrusion processing. The rules are created and installed inside the Neuro-fuzzy designer module in MATLAB platform. There are 625 rules. Each rule consists of 4 criteria to achieve 1 output of solution.

Figure 11 shows the modeling simulation result on the rule viewer. There are 4 input variables as conditions of a test-based problem and a solution. As shown in the example, the degree of extrudability is 80, extrusion ratio is 55, the hydraulic pressure is 725 MPa

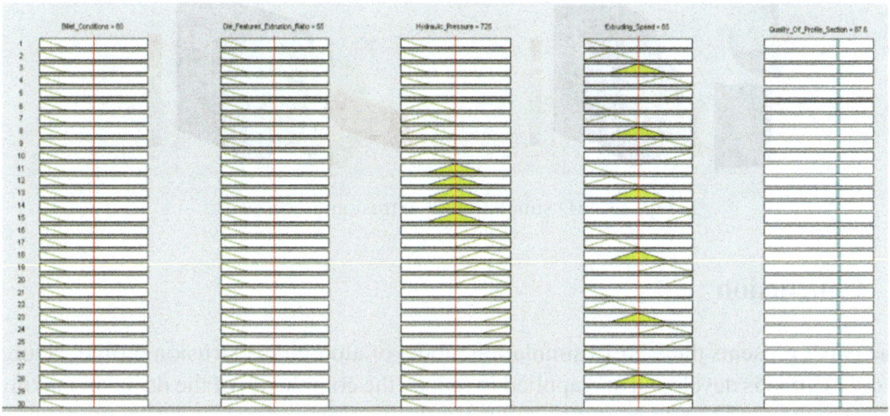

Fig. 11. Rule viewer for the solution of the modeling simulation

and the extruding speed at 65 mm/s, then the profile quality is expected to achieve at 87.5%. In this simulation model, various quality level of the profile can be predicted based on the data input conditions.

5 Results and Discussion

This section explains the results of the modelling simulation using case study. Figure 12 shows the drawing of profile testing and die opening. Figure 13 shows 3D simulation aimed for validation the prediction from the ANFIS modeling. It is found that the ANFIS modeling system can perform similar results to the simulation by using HyperXtrud but validating time is shorter. It is also founded that the left model is better than the right modelling. However, the simulation can illustrate the data of the extrusion variables. For example, the suitable ram speed is 4 mm/s when the AA6063 is tested. The appropriate temperature is 531.75 °C.

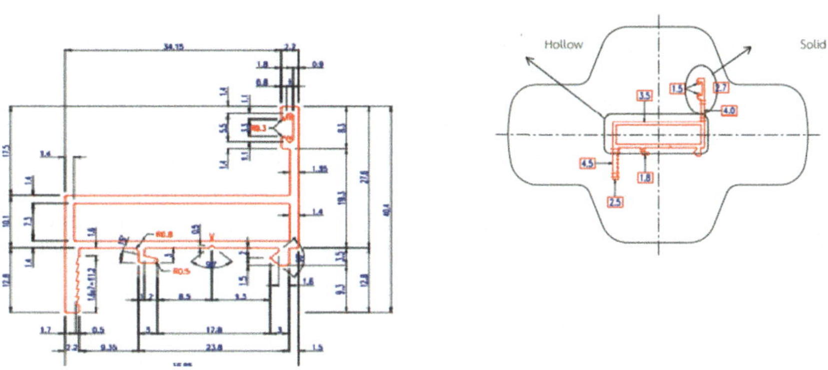

Fig. 12. Drawing of the profile for testing and its die opening

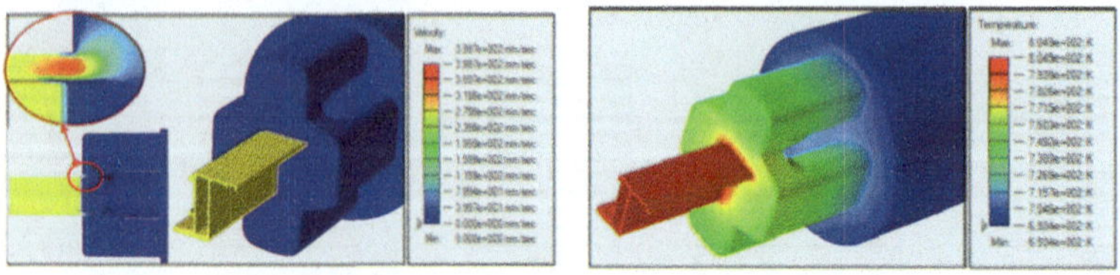

Fig. 13. 3D simulation of extrusion processing

6 Conclusion

The paper presents the ANFIS simulation model of aluminum extrusion profile. Neuro-fuzzy method is developed and applied to reduce the complexity of the decision making. It uses rule based to find solutions and learn from training. The simulation modeling is developed and presented such as the input membership functions, the ANFIS creation, and the output membership function. The modelling is successfully tested. The contribution of the paper is the model development for prediction quality of the profile extrusion beforehand. In addition, the shop floor manager can use the system to determine the optimal and workable conditions for input variables from the given target of the output extrusion profile. However, the research needs to be further developed in the area of learning from experience and test with the physical cases in the factory.

References

1. Pradip, S.: Aluminum Extrusion Technology. ASM International® Materials Park, Ohio (2000)
2. https://www.kaiseraluminum.com/. Accessed 25 Oct 2021
3. https://www.youtube.com/watch?v=Bc7o_sEdX8U
4. Kathirgamanathan, P., Neitzert, T.: Optimization method of extrusion dies with a thin and complex shape. Aust. J. Mech. Eng. **12**(3), 317–326 (2014)
5. Zhao, G., Sun, X., Zhang, C., Chen, H., Guan, Y.: Process modeling and die optimization design of aluminum alloy extrusion profiles used in high-speed train. In: The 11th International Conference on Numerical Methods in Industrial Forming Processes, pp. 144–149 (2013)
6. Kapadia, N., Dasai, A.: Optimization study of die extrusion process using FEM. Int. J. Sci. Res. Dev. JSRD **3**(9), 1007–1009 (2015)
7. Park, N., et al.: Evaluation of the effect of ram speed for extrusion of Al6063 based on ALE-based finite element analysis of L-shaped sample. Procedia Manuf. **50**, 673–676 (2020)
8. Ferras, A.F., Almeida, F., Silva, E.C., Correia, A., Silva, F.J.G.: Scrap production of extruded aluminum alloys by direct extrusion. Procedia Manuf. **38**, 1731–1740 (2019)
9. Suthep, B.: Adaptive aluminum extrusion die design using case-based reasoning and artificial neural networks. In: Advanced Material Research, vol. 383–390, pp. 6747–6754 (2012)
10. Zhang, Y., Zhou, Y., Guo, X., Song, K., Zhang, X.: Fatigue failure prediction model and verification of hot extrusion die. Procedia Manuf. **37**, 66–72 (2019)
11. Negozio, M., Pelaccia, R., Donati, L., Reggiani, B., Tomesani, L., Pinter, T.: FEM validation of front end and back end defects evolution in AA6063 and AA6082 aluminum alloys profiles. Procedia Manuf. **40**, 202–208 (2020)
12. Yadav, R.R., Dewang, Y., Raghuwanshi, J., Sharma, V.: Finite element analysis of extrusion process using aluminum alloy. Mater. Today Proc. **24**, 500–509 (2020)

13. Fang, G., Zhou, J., Duszczyk, J.: FEM simulation of aluminum extrusion through two-hole multi-step pocket dies. J. Mater. Process. Technol. **209**, 1891–1900 (2009)
14. Kniazkin, I., Vlasov, A.: Quality prediction of longitudinal seam welds in aluminum profile extrusion based on simulation. Procedia Manuf. **50**, 433–438 (2020)
15. Valberg, H.S., Lefstad, M., Costa, A.L.M.: On the mechanism of formation of back-end defects in the extrusion process. Procedia Manuf. **47**, 245–252 (2020)
16. Butdee, S., Khanawapee, U.: Quality prediction modeling of a preform fastener process using fuzzy logic and DEFORM simulation. Int. J. Technol. **12**(1), 33–41 (2021)
17. Butdee, S., Phuangsalee, P.: Uncertain risk assessment modelling for bus body manufacturing supply chain using AHP and Fuzzy AHP. Procedia Manuf. **30**, 663–670 (2019)
18. Butdee, S., Noomtong, C., Tichkiewitch, S.: A process planning system with feature based neural network search strategy aluminum extrusion die manufacturing. AIJSTPME **1**(1), 27–53 (2008)

Author Index

A. Hamrol et al. (Eds.): MANUFACTURING 2022, LNME, pp. 251–252, 2022.
https://doi.org/10.1007/978-3-031-00218-2

The manufacturer's authorised representative in the EU is Springer Nature Customer Service Centre GmbH, Europaplatz 3, 69115 Heidelberg, Germany. If you have any concerns regarding our products, please contact ProductSafety@springernature.com

Printed and bound by CPI Group (UK) Ltd, Croydon, CR0 4YY
30/09/2026
02225320-0009